LIFE ON OTHER WORLDS

BY
H. SPENCER JONES

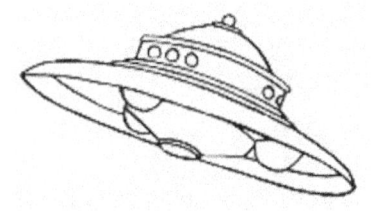

SAUCERIAN PUBLISHER

ISBN: **978-1-955087-28-5**

9 781955 087285

2022, Saucerian Publisher

WORLDS WITHOUT END

BY

THE ASTRONOMER ROYAL

"His charmingly lucid book."

H. G. WELLS

"A coherent survey, concisely and precisely made, of all that any intelligent reader should wish to know."

SIR RICHARD GREGORY

"Written with singular clearness, it gives a most admirably balanced popular survey of current astronomical knowledge and speculation."

THE BISHOP OF BIRMINGHAM

LIFE ON OTHER WORLDS

BY

H. SPENCER JONES
M.A., Sc.D., F.R.S.

ASTRONOMER ROYAL
HONORARY FELLOW, JESUS COLLEGE, CAMBRIDGE

PREFACE

In his Preface to *La Pluralité des Mondes*, Bernard de Fontenelle (1657–1757) wrote:
" I have chosen that part of Philosophy which is most like to excite curiosity; for what can more concern us, than to know how this world which we inhabit, is made; and whether there be any other worlds like it, which are also inhabited as this is ? They who have any thoughts to lose, may throw them away upon such subjects as this; but I suppose they who can spend their time better, will not be at so vain and fruitless an expence." (John Glanvill's translation of 1688.)

The question whether life exists on other worlds is one that still excites curiosity, and to which astronomers are expected to give an answer. In this book I have summarised the evidence, provided by present-day astronomy, which has a bearing on this question, and have attempted to give an answer. I do not expect that every reader will agree with the conclusions that I have reached; it is open to each reader to form his own conclusions, provided that they do not conflict with the evidence derived from the astronomical observations.

The most serious difficulty in discussing whether there is life on other worlds is, naturally, our ignorance of how life originated on the Earth. I have assumed that wherever in the Universe conditions are suitable for life to exist, life will somehow come into existence. The fact that there is pretty conclusive evidence of the presence of vegetation on Mars, the only world where from *a priori* considerations we are able to conclude that conditions are suitable for the existence of life, lends some support to this assumption.   There are those who will not consider this assumption to be justifiable and others who may regard it as opposed to religion.   For the latter, I will quote again from de Fontenelle:

" There remains no more to be said . . . but to a sort of People who perhaps will not be easily satisfy'd; not but that I have good reasons to give 'em, but because the best that can be given, will not content 'em; they are those scrupulous Persons, who imagine, that the placing Inhabitants any where, but upon the Earth, will prove dangerous to Religion: I know how excessively tender some are in Religious Matters, and therefore I am very unwilling to give any offence in what I publish to People, whose opinion is contrary to what I maintain.   But Religion can receive no prejudice by my System, which fills our infinity of worlds with Inhabitants, if a little errour of the Imagination be but rectify'd. . . .   And to think there may be more Worlds than one, is neither against Reason, or Scripture.   If God glorify'd himself in making one World, the more Worlds he made, the greater must be his Glory."

The first chapter presents the picture of the Universe provided by modern astronomy, serving as a background against which the problem under discussion may be viewed in its proper perspective. With the Universe constructed on so vast a scale, it would seem inherently improbable that our small Earth can be the only home of life. In the second chapter the conditions for the existence of life are considered. These conditions are based on the essential uniformity of matter throughout the Universe. The same atoms, subject to the same chemical laws, are found in the Earth and in the remotest star or nebula. In living matter, the chemistry of the carbon atom, intimately bound up with its peculiar property of forming in combination with other atoms a multitude of large and complex molecular aggregations, is of particular significance. The simplest living cells are highly complex, and correspondingly fragile, organisations. The conditions for such complex molecular aggregations to be possible are very restricted. The question whether or not the various planets in the solar system comply with these conditions is discussed in detail in later chapters.

But before embarking on this discussion, an outline of the methods of investigation that are open to the astronomer is given in Chapter III. From purely theoretical considerations it is possible to infer whether a planet is likely to have an atmosphere or not and also to estimate its temperature within fairly narrow limits. These conclusions can be checked by observations. Information about the composition of the atmosphere of a planet can be obtained both from general considerations and from direct observations.

The atmosphere of the Earth and its evolution is discussed in Chapter IV, in relation to the conclusions reached in the preceding chapter. The presence of free oxygen in the atmosphere of the

Earth, of particular importance for the existence of animal life, is attributed to the widespread vegetation over the surface of the Earth.

The detailed discussion of the different members of the solar system shows a wide diversity of conditions; for the most part these conditions are such that the possibility of the existence of life can be definitely excluded. Venus appears, however, to be a world where life may be on the verge of coming into existence and Mars a world where life is in the sere and yellow leaf.

Passing beyond the solar system, the question arises whether other stars can be expected normally to have systems of planets associated with them. In order to answer this question, the origin of the solar system must be explained. This has proved to be a difficult matter. The various theories are discussed in Chapter IX, and it is concluded that, if the distribution of the stars at the time the solar system was formed was very much as it is at the present time, such a special combination of circumstances is needed to account for the solar system that it can be inferred that relatively few of the stars have families of planets.

In the final chapter these conclusions are discussed in relation to the picture of the Universe given in the first chapter. Though conditions are heavily weighted against suitability for the existence of life, it is concluded that there must nevertheless be many other worlds where the appropriate conditions are to be found and where therefore we may suppose that life in some form or other does actually exist.

I am greatly indebted to Dr. W. S. Adams, Director of the Mount Wilson Observatory, California; to Dr. W. H. Wright, Director of the Lick Observatory, Mount Hamilton, California; and to Dr. V. M. Slipher, Director of the Lowell Observatory, Flagstaff, Arizona, for permission to reproduce photographs taken at these observatories and for kindly supplying photographs for this purpose. I am also indebted to the Rev. T. E. R. Phillips for permission to reproduce two of his excellent series of drawings of the planet Jupiter.

H. SPENCER JONES.

CONTENTS

LIST OF PLATES

PLATE I

STAR CLOUD IN THE CONSTELLATION OF SAGITTARIUS (THE ARCHER)

The photograph shows a part of the brightest region of the Milky Way. The denser parts of the cloud are formed by a continuous background of faint stars, which are so numerous that their images blend together. There are numerous dark lanes and channels visible over the cloud, caused by patches of obscuring dust.

The centre of the Milky Way is in the direction of Sagittarius, so that we are here looking through our stellar universe to its greatest depths. It is for this reason that the density of the stars seen in this region is so great.

The distance of the centre of the Milky Way system from the Sun is about 30,000 light-years, or 180,000 million million miles.

Photograph taken by Mr. Franklin Adams.

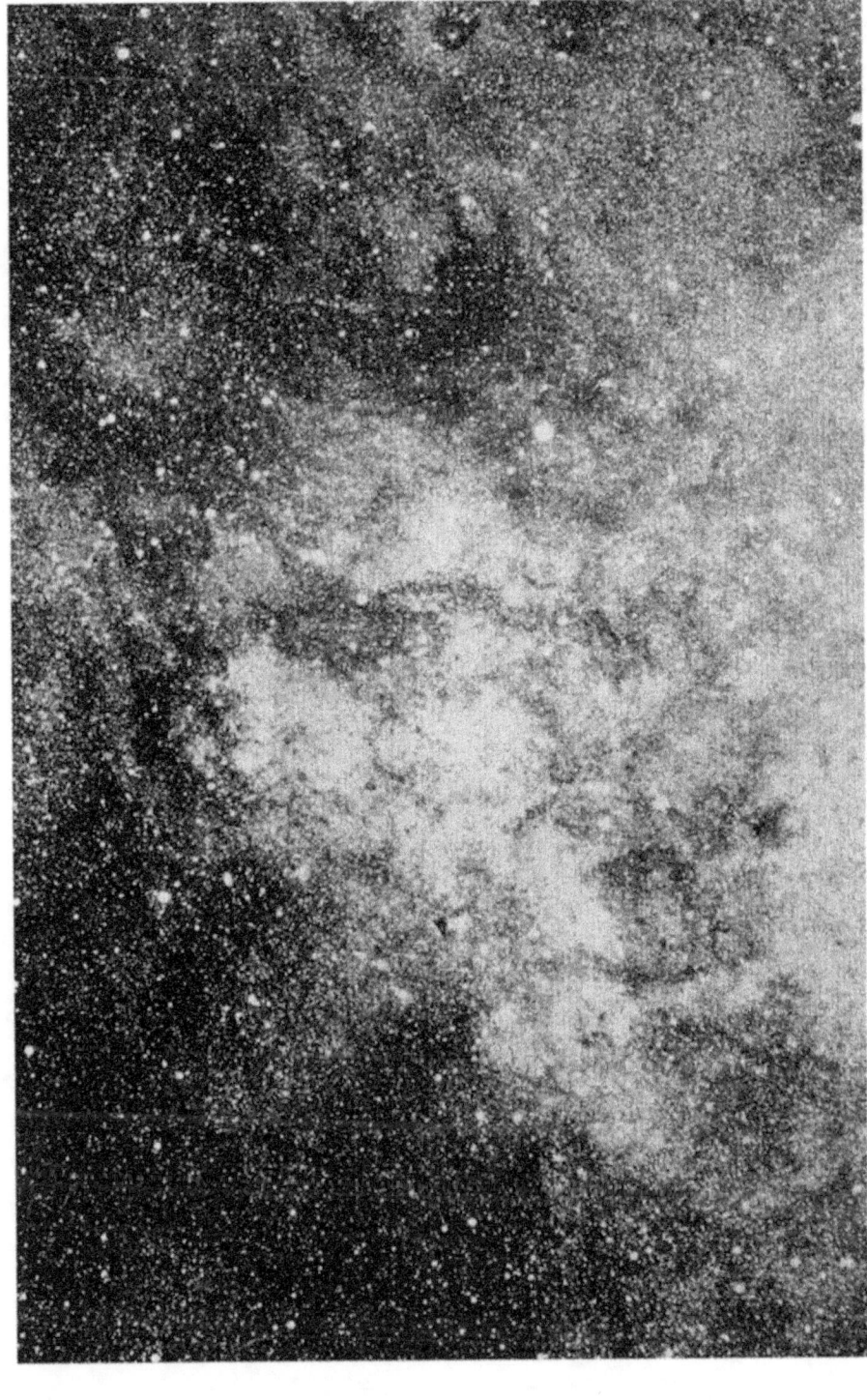

CHAPTER I

A PICTURE OF THE UNIVERSE

THE investigation of the structure of the universe may be said to commence with the work of William Herschel (1738–1822). By profession a musician, Herschel came to England from Hanover at the age of nineteen to improve his fortunes. He earned his living by playing the organ and teaching music, and in 1766 was appointed organist of the Octagon Chapel, Bath. But all his spare time was occupied with the study of optics and astronomy, in which he became more and more interested. Merely to learn about astronomy from books was not sufficient for his eager inquiring mind. He longed to study the heavens through a telescope with his own eyes. Being unable to afford to buy a telescope, he decided to attempt to make one for himself.

Starting in a modest way, Herschel proceeded to larger and larger telescopes. It was not long before he became better at making telescopes than anyone had been before and, what is more, he showed extraordinary skill in using them. To quote his own words: " Seeing is in some respects an art, which must be learnt." It was the excellent optical quality of his telescopes, combined with his keenness of vision, that enabled him to make the discovery that brought him fame. On March 13th, 1781, he noticed what he described in his journal as a " curious either nebulous star or perhaps a comet," an object showing a small round disk, differing in appearance from the point-like image of a star. This object proved to be a new planet, now known as Uranus; it was the first planet ever to have been discovered by man, for the other planets known at that time are easily visible to the naked eye and had been known from remote antiquity. The distance of Uranus from the Sun is twice the distance of Saturn, the remotest of the planets known before Herschel's discovery, so that the discovery doubled the extent of the solar system. It was natural that it should cause a great deal of interest and excitement; one outcome of it was that the King, George III, created for Herschel the post of King's Astronomer, with a salary of £200 a year. This enabled him to give up his musical career for good and all in favour of astronomy.

Free to devote himself entirely to the pursuit of astronomy and the making of telescopes, Herschel set before himself the arduous and gigantic task of making a regular and systematic survey of the entire heavens. He examined each object that came into the field of his telescope and noted any peculiarities about it. He found that the Milky Way, the belt of diffuse whitish light that encircles the heavens, was composed of a vast number of faint stars. By counting the number of stars visible in the field of his telescope in different parts of the sky, he gauged the depths of the stellar universe. From these investigations he came to the conclusion that the universe was an extensive flattened system, shaped very much like a grindstone. He believed that the Sun was somewhere near the centre of this system. The appearance of the Milky Way was the result of looking through the system, from a central position, in the directions of its greatest extension.

In the course of his systematic survey, William Herschel discovered no fewer than 2,500 nebulæ and star clusters. The nebulæ were cloud-like objects, shining with a diffuse light, and he was inclined at first to believe that they were really composed of a multitude of stars so distant that his telescope could not resolve them into separate points of light, just as the naked eye is unable to resolve the Milky Way into discrete stars. But gradually he became convinced that many of them were not aggregations of stars, but masses of glowing gas. This conclusion was confirmed forty-two years after his death, when their light was analysed with the spectroscope and found to show the characteristics of the light from a glowing gas.

At the same time he did not believe that all the nebulæ were gaseous. He was convinced that some of them were composed of a multitude of stars and he regarded these as island universes, comparable in size with our own stellar universe. He said on one occasion: " I have looked farther into space than ever human being did before me. I have observed stars of which the light, it can be proved, must take two million years to reach the earth." This was a remarkable statement to make at a time when the distance of not a single star was known.

These investigations and conclusions of William Herschel have been summarised at some length because they provided for the first time a picture of the universe based on a systematic detailed survey

of the sky.   Subsequent investigation has confirmed
in its essentials the picture painted by Herschel.
Many details have been added and a few of his conclusions have been modified.   Not without reason
has he been called the father of modern astronomy
and well merited were the words in his epitaph:
" Cœlorum perrupit claustra."   He broke through
the barriers of the heavens.

It was not until 1835, thirteen years after Herschel's death, that the distance of a star was
measured for the first time.   The principle of the
method is simple enough; it is essentially the same
as that employed by the surveyor to measure distances on the Earth's surface.   The object whose
distance is required is observed from the two ends of
a base-line of known length, the observations giving
the angle between the directions from the object to
the two ends of the base-line.   The difficulty of
measuring star distances arises from the fact that
the longest base-line available for the purpose is very
short compared with the distances of the stars.   By
making the observations when the Earth is at the
two ends of its orbit, a base-line of about 186,000,000
miles is obtained.   No longer base-line is possible.
To appreciate the problem that faces the surveyor
of the skies, let us represent this base-line of
186,000,000 miles by a line two inches long; the
distance of the nearest star on the same scale is then
about four miles.   We have to measure, in effect,
the distance of an object four miles away by making
observations from two points only two inches apart!
Still more difficult is it when, from the same two
points, we have to measure distances of ten, a

hundred or a thousand miles. Success has been attained by careful attention to detail and by precautions to eliminate as far as possible every source of error.

It is convenient to express star distances in terms of the time that light takes to travel. Light travels with a speed of 186,000 miles a second, so that in the course of a year it will travel a distance of nearly six million million miles. Thus, for instance, instead of saying that the nearest star is 25 million million miles away, we may say that it is about four light-years away. This mode of expressing the distance has the additional interest of reminding us that we see the star, not where it is at the moment, but where it was four years previously.

There is a limit to the distances that can be determined by direct measurement. For distances greater than about 500 light-years, the results become rather uncertain. If space is to be explored to greater distances, it must be by indirect methods. Such a method, which is of very great power, has been discovered, and the knowledge that has been obtained within the last two decades about the structure of the universe has been acquired in a very large measure by the application of this method. It is based on the special properties of a particular class of stars. These stars do not shine with a steady constant light; their light waxes and wanes in a perfectly regular manner. It has been found that the fluctuations in the brightness or candle-power of such a star are accompanied by regular pulsations of the whole star; the star swells up and contracts with perfect regularity. The time required for a single pulsation to be completed, though constant for any one star, ranges for different stars from several hours to about thirty days; if the pulsating stars are arranged in the order of the time of a single pulsation, it is found that they have also been arranged in the order of their candle-power. There is, in fact, a definite relationship between the period of pulsation and the candle-power of the star, so that, if the period is known, the candle-power can be inferred.

There is no great difficulty in finding the time taken by a pulsating star in going through one complete cycle of light-variation; we determine this time, and infer the candle-power, or intrinsic brightness, of the star. We can also measure the apparent brightness of the star, the brightness as seen by the eye. The apparent brightness depends on two quantities—the intrinsic brightness and the distance. If we were to remove a star to twice its present distance, it would appear only one-quarter as bright as before. When both the intrinsic and apparent brightness are known, the distance can easily be inferred.

The longer the period of pulsation, the greater is the candle-power of the star. Thus, for instance, if one pulsation is completed in two days, the candle-power is 260 times that of the Sun; if completed in ten days, the candle-power is 1,700 times that of the Sun; and if completed in thirty-six days, the candle-power is 9,600 times that of the Sun. It will be noticed that the candle-power in each of these examples is very much greater than that of the Sun. The pulsating stars are all, fortunately, intrinsically very bright; they are included in the class of stars called *giant stars*. Their great luminosity makes it possible to see them far away across the vast reaches of space, where mediocre stars like the Sun would be lost to view. It is this fact that renders them so useful in the exploration of space to great distances. For example, suppose we discover in a remote star-cloud that there is one of these pulsating stars whose pulsation is completed in thirty-six days and suppose that this star appears as a star of the eleventh magnitude (or about one-hundredth of the brightness of the faintest star visible to the unaided eye). Such a star would be easily visible in a six-inch telescope. Because the pulsation takes thirty-six days, we know that the candle-power of the star is 9,600 times the candle-power of the Sun. Because it appears as a star of the eleventh magnitude, we can infer that its distance must be 50,000 light-years. When we recall that the limiting distance that can be determined with any approach to accuracy by direct measurement is about 500 light-years, it will be realised what a powerful method for the exploration of space is provided by the pulsating stars.

The general principles that underlie the determination of great distances having been briefly described, the intervening steps can be passed over and we can proceed to summarise the information that has been derived about the stellar universe in which we find ourselves.

The general picture drawn by Herschel, of a vast flattened system, shaped like a grindstone or a thin pocket watch, is confirmed in its essentials, but we now have a far more precise idea of the size and structure of the system. The plane of the Milky Way marks the direction of greatest extent of the system, and it is for this reason, as Herschel realised, that the stars are most numerous in the Milky Way regions. The Milky Way is not uniform in brightness, nor is the distribution of the stars in the Milky Way uniform. They tend to cluster into aggregations or star-clouds. The brightest region of the Milky Way, containing the densest aggregation of stars, is in the constellation of Sagittarius in the southern sky.

Herschel believed that the Sun was near the centre of the system. We now know that in this conclusion he was mistaken and that the Sun occupies a position far out from the centre of the system, though near its median plane. In other words, the Sun is a star in one of the Milky Way star-clouds. The centre of the system, as seen from the Sun, is in the direction of the Sagittarius star-clouds, and it is because in this direction we are looking through the system to its greatest depths that the density of the stars appears so great. A photograph of a portion of this region of the Milky Way is reproduced in the Frontispiece.

In addition to the stars there are also the nebulæ, rarefied clouds of luminous gas, which are found only in or near the Milky Way. One of the most beautiful of these is the Great Nebula in Orion, shown in Plate 2, which is visible to the naked eye as a hazy diffuse patch of light. The nebulæ are not self-luminous; they do not shine by any light of their own. We see them only by virtue of the stars embedded in them; the atoms in the nebulæ absorb the light from the stars and re-emit it in radiations of different wave-lengths. Associated with these bright nebulæ there are also to be found dark nebulæ or obscuring clouds. In the midst of some of the densest star-clouds of the Milky Way there can be seen blank spaces, completely or almost completely devoid of stars. A good example of a blank space is shown in Plate 3. Hundreds of such blank patches are known. They cannot be vacant lanes or channels through the stars, for it is beyond reason to suppose that hundreds of such channels, extending to immense distances, could point directly to the Earth. The blank patches are caused by opaque clouds that lie between us and the stars, which they obscure from our sight. The opaqueness of these clouds is due to the presence of extremely fine dust. Small dust particles, of a size comparable with the wave-length of light, have a very great obscuring power, and if the average amount of dust in the cloud is only one fifty-thousandth of an ounce in each square inch of cross-section, the cloud will be completely opaque, whatever its thickness may be.

Though the luminous nebulæ and the opaque clouds may each occur separately, it is more usual to find the two in close association. It seems that gaseous matter and the larger dust particles are extensively spread throughout the Milky Way regions; where the gaseous matter predominates we see bright nebulosity and, where the dust clouds predominate, the stars that lie behind are hidden. Over a great portion of its extent the Milky Way is divided into two branches; this effect is caused by the existence in the central regions of the Milky Way of obscuring clouds of very great extent. Because the dusty matter is so widespread, we can never hope to see to the bounds of the Milky Way. The gaseous matter of the luminous nebulæ and the dusty matter of the opaque nebulæ can be thought of as the residuum of the diffuse gaseous matter of which our stellar universe formerly consisted. This matter has to a large extent been condensed into stars ; the stars are continually gathering in, by their gravitational attraction, matter from the space surrounding them. They are gradually sweeping space clean, but the sweepers are few in comparison with the vast regions that have to be swept, so that the process is yet far from completion. There are reasons for believing that the total matter not yet condensed into stars is about equal in amount to that contained in all the stars.

The pulsating stars have provided the key by means of which the dimensions of our stellar universe have been determined. It is found that its diameter is about one hundred thousand light-years and the distance from the Sun to the centre of the system is about thirty thousand light-years; the Sun is near the centre of a local clustering of stars, or a star-cloud. The motions of the stars have shown that the system is slowly rotating, under its gravitational attraction. Such a system, consisting of discrete stars and scattered matter, does not rotate as a solid body. When a solid body is in rotation, the motion is more rapid the greater the distance from the centre of rotation; a point on the rim of a rotating wheel, for instance, moves more quickly than a point on the hub. But in the heavens the rule is exactly the opposite ; the nearer to the centre of the rotation, the more rapid is the motion. This can be illustrated by the planets in the solar system; the planet nearest the Sun, Mercury, is moving with a speed of thirty miles a second; the Earth is moving with a speed of eighteen miles a second; Neptune is moving with a speed of only three miles a second.

When the movements of the stars are analysed by statistical methods, it appears that the stars in a certain direction have the most rapid motion on the average, and those in the diametrically opposite direction have the least rapid motion; it is also found that the direction in which the stars have the most rapid motion is the direction to the centre of the system. These results provide certain evidence of rotation. But we can learn much more. We can obtain an estimate of the total amount of matter in the system, because it is the gravitational attraction of this matter that controls the rotation. It is found in this way that the system as a whole has a mass about 160,000 million times the mass of the Sun. In this is included the masses of all the stars, including any stars that may have ceased to shine as well as the mass of all the diffused matter scattered throughout the system. It is not possible to say how many stars there are in the system, but 100,000 million may be taken as a rough estimate, indicating the great scale on which the system is built. The time taken by the Sun to make one complete revolution round the centre is about 225 million years; the stars in the neighbourhood of the Sun have an average speed of about 170 miles a second. Each star has, in addition, its own peculiar motion relative to the group; the motion of the Sun, for instance, relative to the surrounding stars is about thirteen miles a second.

In photographs of the Milky Way star-clouds, the stars appear to be so closely crowded together that it would seem that frequent collisions between them must occur. This appearance is quite deceptive, however. The stars are so far apart that our stellar universe as a whole is comparatively empty. We have seen that the nearest star is 25 million million miles away, and it is therefore clear that the neighbourhood of the Sun is pretty free from stars. It might be thought that the Sun is perhaps in a particularly empty part of the system, but this is not so. The density of stars around the Sun is fairly representative of the system as a whole, except perhaps in the regions close to the centre. Jeans has calculated that an actual collision between two stars can occur on the average only once in 600,000 billion years. This is much greater than the age of the stars, so that we may say that to all intents and purposes collisions between two stars never occur.

If we were to travel with the speed of light through our stellar universe, in the direction away from its centre, we should find after some thousands of years that the stars were becoming less numerous. Some time later we should find only a few scattered outlying members of the system, and at last we should leave these behind and find ourselves in outer space, free from stars. If we continued to travel on, should we come to other stellar universes, or is our Milky Way system the one and only universe? We have seen that William Herschel was convinced that some of the nebulæ that he observed were island universes, at such great distances that his telescopes could not resolve them into discrete stars. The analysis with the spectroscope of the light from such nebulæ lends confirmation to Herschel's views, for the light from these nebulæ does not show the characteristics of the light from a glowing gas, but is more like the light from the stars.

The nebulæ in question are generally called spiral nebulæ, because when seen broadside on they show a characteristic spiral structure. The typical spiral nebula has a bright nucleus from which, at diametrically opposite points, two arms emerge and curl round in the form of a spiral. Such nebulæ are to be found at all angles to the line of sight; some are seen obliquely, when the spiral structure may perhaps be traceable, though not so clearly shown as in those that are seen broadside on. Others are seen edgewise on, and then the spiral structure is not evident; but such nebulæ appear exactly like what we have found our own universe to be—a flattened disk-like system. There is an essential continuity from the nebulæ seen broadside-on to those seen edgewise-on, and we can infer that these latter must also possess the spiral structure. Photographs of spiral nebulæ seen broadside on, inclined to the line of sight, and edgewise, are reproduced in Plates 15, 16 and 17 respectively.

For a hundred years after Herschel's death the

question whether the spiral nebulæ were island universes outside our own universe continued to be debated. It has been only within recent years that the question has been finally settled. The key to the whole question was to find the distances of these nebulæ, because if their distances were known we would at once know whether they were inside or outside our stellar system; we would also know their size and would be able to decide whether they were at all comparable in size with our own system.

The problem was solved when it was found that within some of these nebulæ there were stars which showed all the characteristics of the pulsating stars. The nebulæ in which these stars were found were those of largest apparent size and therefore presumably the nearest to us. Their periods of pulsation were determined and their distances were inferred. They were found to be of the order of a million light-years. This was conclusive evidence that the spiral nebulæ were outside our stellar universe and that they were, in fact, island universes. The close agreement between the distances of the nearer external universes and the estimate of distance given by Herschel (two million light-years) may be noted.

The size of these other universes proves to be of the same general order as that of our own universe. It is found also that they are, like our universe, in slow rotation; they may be thought of as gigantic celestial catherine wheels, spinning round, with their vast spiral arms. They seem also to contain about the same amount of matter as our own system. The nearest of the external universes can be studied in some detail with the aid of the modern powerful telescopes. The typical features of our own system are shown by them—aggregations of stars into star-clouds, bright gaseous nebulosity and opaque dust clouds. The obscuring dust clouds are found to be extensively scattered throughout the central plane of each system, as they are in the region of the Milky Way.

These external universes are of all grades of apparent size from the nearest, which have an angular diameter of a few degrees, to very remote systems, whose images on the photographic plate are scarcely distinguishable from the images of stars. If we make the assumption that the universes are all of much about the same size, we can make rough estimates of their distances. These estimates have served to establish a remarkable fact, which can in turn be used to provide a much more certain determination of the distances. The velocity of each system directly towards us or directly away from us can be measured; the principle used for this purpose is that if a body, which is sending out radiations in the form of light, is approaching, its radiations are slightly compressed together so that their wave-lengths are slightly shorter than they would be if the body were at rest; if, on the other hand, it is moving away, the radiation that we receive are all of slightly longer wave-length.

It is found that the external universes are moving away from us and that the more distant they are, the more rapid is their velocity of recession. This is not the place to discuss the possible interpretations of this remarkable result. One suggestion is that the universes are the fragments of one large

PLATE 2

THE GREAT NEBULA IN ORION

This nebula is visible to the naked eye as a hazy patch in Orion's dagger, below the middle of the three bright stars that form his belt. The bright portion measures about six million million miles across, but the extent of the fainter outer portion is fully three times as great. The nebula is composed of gaseous matter of extremely low density, about one thousand million millionth of that of air under ordinary conditions (much less than that of the most perfect artificial vacuum), yet because of its great extent the total amount of matter in the nebula has a mass about ten thousand times that of the Sun. The nebula glows by the light of stars embedded within it. The round white dots are images of stars (other Suns), the large round patch near the top of the photograph being the image of a bright star.

Photograph by Messrs. Ritchey and Pease, with the 24-inch reflector at the Yerkes Observatory, 1901, October 19.

PLATE 3

THE MILKY WAY NEAR THE STAR RHO OPHIUCHI

The photograph depicts a portion of the Milky Way in the neighbourhood of the star Rho Ophiuchi, not far from the bright naked-eye star Antares in the constellation of the Scorpion. The background of the photograph shows images of a great number of faint stars— faint because of their great distance and not because they are of intrinsically low luminosity.

In the right-hand portion of the photograph is seen a sharply defined lane, almost devoid of star images. This appearance is caused by a cloud of obscuring dust, relatively near to the Sun, which hides the distant stars. The few stars whose images are seen projected on the cloud lie between the Sun and the cloud. Other patches of obscuration may also be seen.

The central region shows patches of bright nebulosity, caused by glowing gas. Obscuration by clouds of dust and bright nebulosity are often found in the Milky Way regions in close association.

Photograph by Dr. E. E. Barnard, with the 10-inch Bruce lens of the Mount Wilson Observatory, 1905, April 5. Exposure $4\frac{1}{2}$ hours.

universe, which was originally very compact.   An explosion occurred and the fragments were sent flying through space in all directions.   If such were the case then, after the lapse of a considerable time, we would find that the most rapidly moving parts would be at the greatest distances and that, viewed from any one of the fragments, all the other fragments would appear to be moving away and with speeds that were faster the greater the distances. This suggested explanation may not be the true one, but it serves to illustrate that there may be a simple interpretation to facts that at first sight appear very strange.

The point of interest in the present connection is that the close proportionality between velocity of recession and distance affords by far the most accurate and reliable method of estimating the distances of very remote systems, because the velocity in the line-of-sight can be estimated from the wavelengths of the radiations with good accuracy.   It is found that the velocity in miles a second divided by 106 gives the distance in millions of light-years. Thus, for instance, a distant universe in the constellation of Boötes has been found to be receding with a velocity of 24,300 miles a second.   We can infer that this nebula is at a distance of about 230 million light-years.   Such a distance is beyond the power of the imagination to conceive.   Whilst the light by which this universe is seen has been travelling through space, the dinosaurs and flying reptiles have appeared on the Earth and with the slow march of evolution have disappeared again. Mountain ranges have been uplifted and then worn down by erosion. The surface of the Earth has entirely changed its appearance. And at length, when the light was nearing the end of its long journey, man appeared on the Earth. Such great distances, though they may surpass our powers of conception, cannot fail to impress upon us the vastness of space.

The most distant systems that have been recorded on long-exposure photographs with the great 100-inch telescope are at a distance of about 500 million light-years. Within a sphere of this radius, it is estimated that there are about one hundred million universes, the average distance of any universe from its nearest neighbour being of the order of a million light-years. At these greatest distances to which space has as yet been probed, the universes seem to be scattered with an approximately uniform distribution; there is no evidence of any falling off in density at these extreme distances nor of any approach to the bounds of space.

Such, in brief, is the picture of the universe provided by modern astronomical observation. We see the Earth as a small planet, one member of a family of planets revolving round the Sun; the Sun, in turn, is an average star situated somewhat far out from the centre of a vast system, in which the stars are numbered by many thousands of millions; there are many millions of such systems, more or less similar to each other, peopling space to the farthest limits to which modern exploration has reached.

Can it be that throughout the vast deeps of space nowhere but on our own little Earth is life to be found ? Can astronomy tell us whether life can

exist on any of the other planets belonging to the
solar system; and if it can exist, whether it does
exist ? Is it possible to estimate the likelihood of
life existing somewhere in the universe outside the
solar system ? These are some of the questions
that I am continually being asked. In succeeding
chapters the attempt will be made to answer these
questions, in so far as it is possible for astronomy to
provide answers. But first we must consider what
life is and what tests can be used to decide whether
life is possible or is not possible on any particular
world.

CHAPTER II

THE CONDITIONS FOR THE EXISTENCE OF LIFE

IN attempting to discuss whether life can exist on any other world, we come up against the difficulty that we have no certain knowledge of how life originated on the Earth. Suppose we could show that on some other world the conditions were essentially similar to those on the Earth. Would it be legitimate to assume that because life has come into existence on the Earth, there must necessarily be life also on the other world, though perhaps in different forms from those with which we are familiar ? If, on the other hand, we could show that the conditions on another world differed from those on the Earth to such an extent that no forms of life now present on the Earth could exist, would it be a legitimate conclusion that the other world must be a world devoid of life ? May we not have some justification for assuming that the forms of life that now exist on the Earth have developed, through a slow process of evolution, to suit those conditions and that, if different conditions were found to prevail elsewhere in the universe, different forms of life might have evolved ? It is conceivable, for instance, that we could have beings, the cells of whose bodies contained silicon, instead of the carbon which is an essential constituent of our cells and of all other living cells on the Earth ; and that, because of this essential difference between the constitution of these cells and the cells of which animal and plant life on the Earth are built up, they might be able to exist at temperatures so high that no terrestrial types of life could survive. To obtain some guidance in endeavouring to answer such questions, we must consider what biology can teach us about the nature of life.

All forms of matter, inorganic, organic or living matter, are built up of the atoms of different elements. Ninety-two different elements are known to the chemist, hydrogen being the lightest and uranium the heaviest. The atoms of these elements may be regarded as the bricks from which all matter, everywhere in the universe, is built up. It might seem at first sight surprising that the great variety of substances that we find on the Earth can be built from so limited a number of different atoms. But the variety is produced by the great variety of ways in which the different types of atoms can be combined, just as the richness of our language is the result of the large number of ways in which the twenty-six letters of the alphabet can be combined to form words.

So all the innumerable substances that we find on the Earth or that we can think of—the minerals in the crust, precious stones, bricks, timber, all living things, coal, oil and so forth—are merely the result of different combinations of some of the ninety-two varieties of atoms. But these same atoms are found also in the Sun, in the stars—the most distant as well as the nearest—in the nebulæ and in the remote universe, as well as in the diffuse gaseous matter and star-dust that is thinly scattered about space between the stars. Conversely, no element is known to occur in the Sun or the stars that has not been found on the Earth. It is true that one element was discovered in the Sun before it had been found on the Earth; that was the gaseous substance to which the name *helium* (from the Greek word for the Sun, ἥλιος) was given. Helium was subsequently discovered on the Earth in cleveite and other minerals containing uranium and was found to be present in small quantities even in the air that we breathe.

But not merely do we find in the Sun and the stars the same elements that are found on the Earth, we find also that the elements that are the most abundant on the Earth are on the whole the most abundant in the Sun and the stars, and those that are least abundant on the Earth are the least abundant in the Sun and the stars. There are some exceptions, which are not without significance as we shall see subsequently, but on the whole the parallelism is very striking and suggests a common origin from some primordial matter.

How can we detect the presence of this or that particular element in the Sun or in a distant star? The detection is made by analysing the light with a spectroscope, which breaks the light up into its different constituents. The light reaching us from the Sun is highly complex; the atoms of any given element can vibrate in a number of different ways and each particular mode of vibration gives rise to the emission of light of one particular wave-length. At any instant, some of the atoms will be vibrating in one particular mode, others in a different mode and so on. The aggregate of the light radiations corresponding to these various vibrations of the atom gives what is called the *spectrum* of the element, a series of light-vibrations of definite wave-lengths or frequencies which is characteristic of the element in question and is produced by no other element. If this particular series of vibrations is detected when the light from a star is analysed, then we may conclude with absolute certainty that the element in question is present in the star.

The atoms were formerly thought to be small, hard spheres, the smallest particles of a substance that could exist by themselves. If an element were divided into smaller and smaller particles, we should at length come to a stage when we could subdivide no further. We should then have reached the individual atoms. But modern investigations of the structure of matter have shown that this picture is incomplete. The different atoms are themselves built up of elementary particles, called protons, neutrons and electrons. The proton carries a positive charge of electricity; the electron carries an equal negative charge; the neutron, as its name implies, is electrically neutral. The proton and the neutron have approximately the same mass and are much heavier than the electron. A fourth elementary particle, which has been called the positron or positive electron, has recently been discovered; it has the same mass as the negatively charged electron but a positive charge equal to that of the proton. The positron is not usually detected in the presence of matter; it is possible, though not certain, that the proton is merely a combination of a neutron and a positron, in which case there would be three fundamental particles-neutrons, negative electrons and positive electrons. The atom as a whole is electrically neutral, so that the total number of protons contained in it is equal to the total number of negative electrons.

The modern conception of an atom is of a system containing a nucleus, in which almost all the mass of the atom is concentrated, composed of protons and neutrons, whilst outside the nucleus there are sufficient electrons to make the atom electrically neutral; we may think of the electrons as describing orbits around the nucleus, but the system as a whole is much more complicated than a miniature solar system. In the solar system, each planet moves in a definite orbit around the sun; but in the atom, each electron can move in a number of different orbits and can jump from one orbit to another.

The simplest atom is the atom of hydrogen, containing a nucleus of a single proton with a single electron outside it. It is also necessarily the lightest of the atoms. The next simplest atom is the atom of helium; the nucleus of the helium atom contains two protons and two neutrons and there are two external electrons. The weight of this atom is approximately four times the weight of the atom of hydrogen or, in other words, we say that helium has an atomic weight of four. So, in succession, the atoms of the different elements can be built up, each containing one more external electron than the preceding and the atomic weight being determined by the number of protons and neutrons in the nucleus. It will now be readily understood why the same atoms are to be found everywhere in the universe and why we do not find in the Sun, for instance, a series of atoms entirely distinct from those that are found on the Earth. The atoms throughout the universe are built up of the same fundamental particles, and out of these we can build up one and only one series of atoms, which increase in complexity from the lightest elements to the heaviest.

The atom is something very different from the hard solid sphere that it was formerly believed to be. Yet in ordinary chemical processes the atoms of the different elements retain their identity. The atom of helium does not split up and form four atoms of hydrogen. The nucleus of the atom must be split up in order that we may succeed in transmuting one element into another. This requires a great deal of energy, much more than can be furnished by any chemical action; the splitting up of atoms cannot be brought about, therefore, by chemical action. So although the various atoms are all built up of protons, neutrons and electrons, we may still for most purposes think of the atoms as they were pictured before the new knowledge about the structure of matter had been gained.

The same atoms, with the same structure, being found throughout the universe, it necessarily follows that the chemical laws that they obey will be the same everywhere, because these laws are the result of the particular structures and their energy relationships. The same chemical compounds can exist under the same conditions anywhere in the universe. Thus, for instance, two atoms of hydrogen and one atom of oxygen can unite to form a stable chemical structure, a molecule of water, and one molecule of water is precisely similar to every other molecule. The molecule of a substance is the smallest part of the substance that can exist separately. We shall not find that elsewhere in the universe it will be necessary to have three or four atoms of hydrogen to combine with one atom of oxygen to form a stable chemical compound. This conclusion is of importance for our consideration of the possibility that living matter may take essentially different basic forms elsewhere in the universe from the forms that are found on the Earth. Though types of vegetable or animal life that are unlike any types to be found on the Earth may conceivably occur elsewhere in the universe, the chemical compounds of which the individual cells are made up must be such as could exist on the Earth and it is unlikely that they differ from the compounds of which living matter on the Earth is built up.

A special rôle is played in living organisms by carbon, because it possesses to a far greater extent than any other element the power of uniting with itself, as well as with other elements, to build up single molecules containing very large numbers of atoms. It is these complex molecules containing carbon that form the basis of the structure of all living organisms. The only other element that possesses the power of building up complex molecules to any great degree is silicon, but the compound molecules which have carbon as a basis are far more numerous and complex than those which have silicon as a basis.

Too long a digression into chemical theory would be necessary to explain why carbon possesses this unique property. But it is primarily due to the fact that the atom of carbon is what is called a tetra-valent atom. Into the formation of every molecule of any given chemical compound a definite number of atoms enter. If we consider simple compounds involving hydrogen, we find that one atom of certain elements will combine with one atom of hydrogen; one atom of certain other elements will combine with two atoms of hydrogen; while one atom of yet other elements will combine with three or with four atoms of hydrogen. Thus, for instance, one atom of hydrogen will combine with one atom of the yellowish poison gas chlorine to form hydrochloric acid; chlorine and hydrogen are called monovalent elements. Two atoms of hydrogen will combine with one atom of oxygen to form water; oxygen is called a divalent element. Three atoms of hydrogen will combine with one atom of nitrogen to form the pungent gas ammonia; nitrogen is called a trivalent element. Four atoms of hydrogen will combine with one atom of carbon to form the gas methane, familiar to the miner as the dangerous fire-damp; carbon is called a tetravalent element.

Two elements that unite with each other, one to one, have the same valency. Thus, for instance, the molecule of common salt contains one atom of sodium and one atom of chlorine. Chlorine is monovalent and, therefore, sodium must also be monovalent. The atom of a divalent element may unite with two atoms of a monovalent element, with one atom of each of two monovalent elements, or with one atom of another divalent element. The molecule of caustic soda, for instance, consists of one atom of divalent oxygen linked on one side with one atom of monovalent sodium and on the other side with one atom of monovalent hydrogen; whilst the calcium oxide molecule consists of one atom of calcium and one atom of oxygen, and because oxygen is divalent so also must calcium be.

It will be clear that complex molecules cannot be built up from monovalent atoms; when two such atoms unite with each other there are no free affinities or links left over to which other atoms can attach themselves. The possibilities are greater but are still very limited when we consider divalent elements. The more linkages or affinities that the atom has, the greater are the varieties of molecules that can be built up.

This statement is subject to limitation, however. We have not referred as yet to valencies higher than four. There are some elements, such as phosphorus and nitrogen, which are pentavalent, and it might be thought that these elements would have a greater power of building up complex molecules than carbon has. Such elements, however, tend to behave as trivalent elements, two of the links joining together and cancelling each other; we saw above that in ammonia nitrogen behaves as though it were trivalent. The maximum power of uniting with other atoms seems to be reached with the tetravalent atoms and, amongst the tetravalent atoms, to the largest degree with carbon. The carbon atom can have four linkages joined up with other atoms, when it is said to be saturated, but it is possible also for two of the four to interplay with each other, in which case the carbon atom is said to be unsaturated. An example of an unsaturated carbon atom is provided by carbon monoxide, the poisonous constituent in coal gas or in the exhaust gases from a motor-car engine; the molecule of carbon monoxide contains one atom of carbon and one atom of oxygen.

We will consider a few simple compounds of carbon to illustrate the facility with which a variety of compounds can be built up. We start with methane or marsh-gas, and suppose the atoms of hydrogen are replaced one by one by atoms of monovalent chlorine. Methane and the four compounds that can thus be formed may be represented thus:

$$
\begin{array}{cccc}
\mathrm{H} & \mathrm{Cl} & \mathrm{Cl} & \mathrm{Cl} \\
| & | & | & | \\
\mathrm{H-C-H} & \mathrm{H-C-H} & \mathrm{Cl-C-H} & \mathrm{Cl-C-H} \\
| & | & | & | \\
\mathrm{H} & \mathrm{H} & \mathrm{H} & \mathrm{Cl}
\end{array}
$$

$$
\begin{array}{c}
\mathrm{Cl} \\
| \\
\mathrm{Cl-C-Cl} \\
| \\
\mathrm{Cl}
\end{array}
$$

(H denotes an atom of hydrogen, C an atom of carbon, Cl an atom of chlorine.)
or by the formulæ

$$CH_4, \quad CH_3Cl, \quad CH_2Cl_2, \quad CHCl_3, \quad CCl_4$$

In the first representation, by what chemists term a structural formula, the relationship of the several atoms to one another in each molecule of the substance is shown. The second group of formulæ merely indicates the chemical constitution of the molecules, without giving any information about their structure. The structural formulæ, giving an indication of the structure of the molecules, are much the more informative.

The first substitution of one atom of chlorine gives the substance called methyl chloride. The group (CH_3), in which the carbon atom has one free link, behaves like a simple monovalent atom and can take the place of monovalent atoms in other compounds, so forming compounds of greater complexity; it is called the methyl group. The next compound is di-chlor-methane. By the substitution of a further atom of chlorine we obtain tri-chlor-methane, which is the chemical name for the anæsthetic, chloroform. When all the hydrogen has been replaced by chlorine we obtain carbon tetra-chloride, an important organic solvent, particularly of fatty or greasy substances, which has the advantage of not being highly inflammable. It is therefore used extensively in dry cleaning processes, being safer than inflammable solvents such as petrol or benzene.

The organic substances that are built up by living plants and animals can be classified into three broad divisions, the carbohydrates, the fats and the proteins. We will illustrate briefly how such substances can be built up in progressively increasing complexity. We start with the simplest of the

PLATE 4

THE MOON: THE APENNINES AND ARCHIMEDES

The Apennine Mountains form the greatest range of mountains on the Moon, being nearly 640 miles in length and reaching a maximum height of 21,000 feet. They rise gradually on the S.W. side, but fall away steeply, with great precipices, on the N.W. (In the plate, North is at the top and West at the right.)

North and east of the Apennines lies the great lunar plain, called *Mare Imbrium*, or Sea of Storms. Ripple marks on this plain mark the limits of successive flows of lava.

In the centre of the upper portion of the photograph is the ring-mountain or lunar crater, known as Archimedes. It is 50 miles across. The highest part of the mountain-ring is 7,400 feet above the interior, whose level is 650 feet below that of the surrounding surface.

The two craters near the top right-hand corner are known as Aristillus (upper) and Autolycus (lower).

Photographed by Dr. J. H. Moore and Mr. J. F. Chappell, with the 36-inch refractor, Lick Observatory, 1937, October 26.

PLATE 5

THE MOON: REGION OF TYCHO AND CLAVIUS

The photograph shows part of the southern portion of the Moon's surface, illustrating its extreme ruggedness. The whole of this region is honeycombed with craters of all sizes—from the smallest to the largest, many craters being contained within, or overlapping, the walls of other craters.

Immediately to the left of the centre of the photograph is Tycho, the most perfect specimen of a lunar crater. It is 54 miles wide and 17,000 feet deep, so that Mt. Blanc, if placed inside it, would not reach the top. The central peak, whose shadow can be seen on the floor of the crater, is 5,500 feet in height. From this crater extends a great ray system, which can be seen near full moon; some of the rays extend over thousands of miles, passing across mountains and valleys.

The large crater at the bottom of the photograph is Clavius. It has a diameter of 142 miles. There are numerous small craters on and within its ring.

Photographed by Dr. J. H. Moore and Mr. J. F. Chappell, with the 36-inch refractor, Lick Observatory, 1937, October 26.

sugars, whose basis is six carbon atoms in the form of a chain, with links to hydrogen and to the group called hydroxyl, consisting of an atom of oxygen and an atom of hydrogen (—O—H), which in a similar way to the methyl group acts as a mono-valent atom. The diagrammatic or structural formula is:

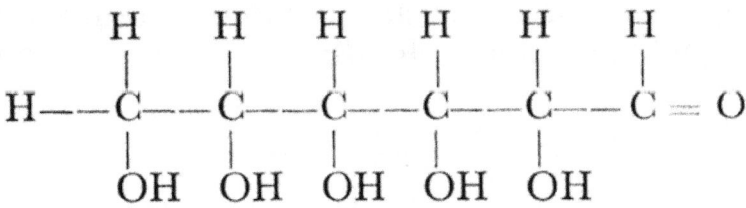

It may be noted in passing that groups containing six carbon atoms occur very frequently in carbon compounds and seem to possess great stability.

A more complex sugar can be obtained by com-bining two such molecules. If we imagine one of the hydrogen atoms removed from the first molecule and one of the hydroxyl groups removed from the second, we can join up the resulting free linkage to form cane sugar, a disaccharide, whilst the hydroxyl group and the hydrogen atom, which have been set free, join up to form a molecule of water, H—O—H. This process can be repeated and, theoretically, we can build up in this way molecules of sugars or starches of any desired degree of complexity. The starches found in living plants contain thirty to forty of such groups. There should in practice presumably be a limit, conditioned by the energy relationships, beyond which further reduplication would not be possible, the molecules becoming unstable and breaking down. But the process illustrates the adaptability of the carbon atom as the basis for large molecular aggregations. The sugars and starches, with this general type of structure, form the carbohydrates of living matter.

The basic structure of fats of living cells can be explained by starting with glycerine; the molecules of this substance consist of a chain of three carbon atoms in which each carbon atom is united to a hydrogen atom and a hydroxyl group, giving the structure:

$$
\begin{array}{ccccccc}
& & H & & H & & H \\
& & | & & | & & | \\
H & \text{---} & C & \text{---} & C & \text{---} & C & \text{---} & H \\
& & | & & | & & | \\
& & OH & & OH & & OH
\end{array}
$$

When each hydroxyl group is replaced by a molecule of a fatty acid, a fat is formed. A hydrogen atom from the fatty acid molecule combines in this process with the hydroxyl group to form water.

The fatty acid is usually built up on a basis of chains of six carbon atoms; the following structural formula represents a simple fatty acid:

$$
\begin{array}{cccccc}
H & H & H & H & H & \\
| & | & | & | & | & \\
H\text{---}C\text{---}C\text{---}C\text{---}C\text{---}C\text{---}C & = & O \\
| & | & | & | & | & | \\
H & H & H & H & H & OH
\end{array}
$$

This formula may be compared with the formula of the simple sugar given above. It is the group

on the right-hand side $-C\!\!\!\begin{array}{c}\diagup O\\ \diagdown OH\end{array}$, called the *carboxyl*
group, which confers the acid properties.

If such a molecule combines with the glycerine molecule, the hydrogen of the (OH) acid group combines with one of the (OH) groups in the glycerine molecule to form water, the two free linkages being joined. When the three (OH) groups in the glycerine molecule have been replaced in this way by three fatty acid molecules, each of which loses the hydrogen atom of its acidic group, a fat is obtained. The fatty acids that are formed in nature in living cells are usually of a more complex nature than this, being derived from the union of three of the six-carbon chains, only one of which possesses the acid carboxyl group. There is then a row of eighteen carbon atoms. Three such complex molecules react with the glycerine molecules to form the large molecule of a fat. It will be realised that the process of carbon-chain building can be considerably extended with the formation of large and very complex molecules, all of which possess, however, the same general characteristics.

To explain the basis of the structure of the third group of substances found in living matter, the proteins, we again start with the simple fatty acid, based on the six-carbon chain, whose formula was given above. If the hydrogen atom at the end of the chain is replaced by the *amino* group, $-N\!\!\!\begin{array}{c}\diagup H\\ \diagdown H\end{array}$, obtained from an ammonia molecule that has lost one of its atoms of hydrogen, and which behaves as a monovalent atom, a substance is obtained that has the formula:

$$\begin{array}{ccccccc}
H & H & H & H & H & H & \\
| & | & | & | & | & | & \\
N\!-\!C\!-\!C\!-\!C\!-\!C\!-\!C\!-\!C & = O \\
| & | & | & | & | & | & | \\
H & H & H & H & H & H & OH
\end{array}$$

(N denotes an atom of nitrogen)

Such a substance is one of a large group of substances that are called amino acids. It will be noted that it contains at one end the carboxyl group, which confers acid properties, whilst at the other end it contains the amino group, which confers alkaline or basic properties. It may be mentioned that ammonia, from which the amino group is derived, is a strong base. The amino acids are, therefore, acid in one part and basic in another and according to the circumstances they can act either as an acid or as a base. The amino acids form the basis of all the proteins. Many of these have an extremely complex structure, with molecular weights running into many thousands. The acid portion of one amino acid has a chemical attraction for the basic portion of another and this enables them to unite to form a more complex amino acid. The hydrogen atoms may be replaced by many diverse groups, of a more or less complicated nature, without affecting the characteristic amino-acid properties. There may be two amino groups instead of one, forming a di-amino acid; sometimes also there are two acid groups. The peculiar property of the carbon atom of forming chains makes possible a vast number of different combinations, with the result that an almost infinite variety of structures can be obtained from the four elements hydrogen, carbon, nitrogen and oxygen.

An important process in nature is the building up of carbohydrate into fat in living structures; the process is important because the fat provides a larger store of energy.   The energy that is stored in this way is derived from sunlight.   Reference to the structural formulæ given above shows that the carbohydrates contain the repeated structure

H   H
| |
—C— , whilst the fats contain the repeated   —C—
| structure   .   .   .   . |
OH   H

The action of sunlight is to split up one of the OH groups in the carbohydrate and to replace it by a hydrogen atom, the oxygen being given off in the process and energy from the sunlight being utilised. This is essentially the process that the chemist calls reduction.   It can be reversed; the organic sub-stance then becomes oxidised and energy is released, which is available for vital processes.

We have seen that the large molecules built up around long chains of carbon atoms form the basis of living matter.   A further stage of complexity is produced by the tendency for a number of these large molecules to unite to form molecular aggre-gations.   In such molecular unions, each molecule behaves as a single atom.

The separate molecules in the aggregation are feebly held together without true atomic union;

very little evolution of energy is involved in the formation of such aggregations, which are in a somewhat unstable condition. They exist in a state of delicate balance. This is the distinguishing feature of the state of matter that is termed the *colloidal* state. The study of matter in this state has become an important branch of chemistry. It has proved of particular significance in biochemistry— the chemistry of living matter—because much of the material of which living cells are constructed exists in this state. The colloidal substances may be either active in solution, or they may be inactive masses which have been formed from the dissolved living colloids and thrown out of solution. The latter process gives rise to the membranes that surround the separate cells and divide one from another. It may be illustrated by the formation of a skin on the surface of warm milk. The chief colloid of the milk accumulates on the surface and the molecular aggregations join together to form a close network or film.

To consider the special properties of matter in the colloidal state and to discuss the reasons why the colloidal state plays so large a part in vital processes would be outside the scope of this book. The important point to emphasise for our present purpose is the tendency for the occurrence in living matter of large molecular units in a state of a delicate balance of equilibrium.

The preceding discussion may be summed up as follows: the same atoms that occur on the Earth are to be found in the remotest parts of the Universe. The same chemical laws necessarily prevail throughout the Universe. The great variety of substances that are necessary to form living matter is made possible by the peculiar power of the carbon atom of uniting with other atoms. The chemistry of carbon is, therefore, of great importance in the study of living matter and wherever in the Universe living matter may occur it must be dependent upon the special properties of the carbon atoms. The formation of large molecules with a chainlike structure and of feebly stable molecular groups must be possible, it would seem, if living matter is to exist.

These conclusions can serve as a guide in considering the conditions that are necessary for the existence of life to be possible. The first requisite is that the temperature should be neither too high nor too low. Every chemical compound can be split up or dissociated by raising its temperature sufficiently. The ease with which various substances can be broken up by heating differs considerably. In the hottest stars no compounds at all can be detected; matter can exist only in the atomic state. In the Sun, whose temperature is about 6,000° C., a few very simple compounds that strongly resist dissociation are found, such as silicon fluoride and cyanogen. The temperature of sunspots is about a thousand degrees lower than the temperature of the rest of the Sun and this lower temperature enables other simple compounds to be present in the spots, which cannot exist at the higher temperature of the rest of the Sun; amongst such compounds are titanium oxide, boron oxide and the hydrides of magnesium and calcium. For the cool red stars, with temperatures of the order of 3,000° C., bands due to compounds become more prominent in the spectra; but the compounds are simple compounds, titanium and zirconium oxides, cyanogen and a few other simple carbon compounds.

In general, the more complex the structure of the molecules of a substance, the more readily they are broken up when the temperature is raised. We have seen that the molecules of which living matter is composed are extremely complex. For this reason they are also very fragile and have but small power to withstand being broken up when the temperature is increased. All forms of life that we know are, in fact, very sensitive to high temperature. The surest way to kill any form of life is to subject it to a high temperature. For this reason milk is sterilised by the process of pasteurisation; no milk can be sold as pasteurised unless it has been maintained at a temperature between 145° F. and 150° F. for not less than 30 minutes. It is found that this treatment ensures the destruction of any disease-producing organisms, such as the tuberculosis bacilli, which may be present in the milk. For the same reasons, water of doubtful purity can be made safe for drinking by being well boiled. Bearing in mind that living matter, wherever it may occur in the Universe, must be complex in structure, we may reasonably conclude that wherever temperatures are sufficiently high to destroy any form of life that occurs on the Earth there can be little expectation that life will be found. We may note, moreover, that the higher forms of life are less resistant to heat than the simple organisms. The temperature might be such that the simplest types of life could exist, though any development of higher and more complex types of life could not occur. Without attempting to lay down any hard and fast limit, it does not seem probable that complex living structures can be expected wherever the temperature is much in excess of about 150° F.

Though most forms of life are unable to survive at very low temperatures, it is known that some can withstand extreme cold for long periods. Low temperatures, unlike high temperatures, do not break up chemical compounds. But when life is not actually destroyed by very low temperatures it becomes latent, as it were. There seems to be a state of suspended animation, in which all vital processes are suspended until the temperature is raised. We cannot believe that, if these conditions prevailed on any other world, life could develop, for how can there be any development if vital processes are suspended ? The reason why low temperatures are inimical to life is evident. The development of vital processes requires energy. On the Earth, the energy of all living things is ultimately dependent upon green plants, which in turn are dependent for their energy upon radiation from the Sun. On a planet in some other solar system, vital processes must similarly be ultimately dependent upon energy received from the parent Sun. The temperature of any such world is conditioned by the energy that it receives from its parent Sun; a low temperature implies that it receives but a small supply of energy. If the temperature is so low that there is not sufficient energy available for vital processes, there can be no life. We must conclude, therefore, that if on any world we find either a high or a low temperature, it is extremely improbable that life in any form could exist on it.

It will be clear that the restriction to a moderate temperature enables all the stars to be at once ruled out of consideration as possible homes of life. For the temperatures of the stars are so high that even on the coolest of them only a few of the simplest compounds are to be found. All the complicated molecules that are present in every type of living organism are broken up by heat at temperatures far below those that prevail in the coolest stars. Our search for life elsewhere in the universe must, therefore, be restricted to the planetary bodies, which have much lower temperatures than the stars. It may be mentioned that, on this subject, Sir William Herschel, whose conclusions about the structure of the Universe were far in advance of his time, held views that were strangely erroneous. He believed that the Sun was a cold, dark body, covered with a layer of fiery clouds. He thought that it was " most probably also inhabited, like the rest of the planets, by beings whose organs are adapted to the peculiar circumstances of that vast globe."

Given satisfactory conditions of temperature, some further clues as to the requirements for the existence of plant or animal life can be obtained from consideration of the requirements for life on the Earth. The green plant builds up carbohydrates in the following way: carbon dioxide is absorbed from the air by the plant and enters into union with he water in the plant, forming carbonic acid

($H_2O + CO_2$), which may be represented by the

formula $HO-\overset{\overset{\textstyle OH}{\textstyle |}}{C} = O$. Under the action of sunlight, one of the OH groups is split up and replaced by hydrogen, the oxygen being given off by the plant to the air. Sunlight provides the energy that is needed for this transformation to be able to take place, but it is the green colouring matter in the plant, called chlorophyll, which makes the transformation possible. For this reason, chlorophyll is called a photocatalyst; it enables the transformation to occur under the action of light, taking some complex part itself in the transformation, though it remains unchanged when the transformation is completed.

By this action, a substance with the formula $HO-\overset{\overset{\textstyle H}{\textstyle |}}{C} = O$, known as formic acid, is produced; it contains the characteristic carbohydrate grouping, $H-C-OH$. When the formic acid is in turn reduced, and six of the resulting groups unite to form a six-carbon chain, a carbohydrate is obtained; the formation of more complex carbohydrates and of fats, in the way already explained, follows naturally.

There are many lower plants that obtain all their carbon in the dark, without the action of sunlight, by the reduction of carbon dioxide and then synthesise their organic constituents from that source. In such cases, chemical energy provides the supply of energy that is needed, instead of the luminous energy from sunlight. Whichever process is followed, a supply of carbon dioxide is essential, from which the plant can obtain its carbon. It would seem, therefore, that plant life is dependent upon carbon dioxide being available, though the carbon dioxide need not necessarily be present in large amount. The proportion of carbon dioxide contained in the atmosphere of the Earth is only between three and four parts in ten thousand; this small proportion supplies the needs of the extensive plant life on the Earth.

The supply of carbon dioxide in the Earth's atmosphere is maintained by the process of combustion. The combustion of coal consists essentially in the combination of carbon and oxygen, with the production of carbon dioxide. If a supply of oxygen is not available the coal will not burn. The process of burning coal, considered as a chemical action, is, therefore, merely the oxidation of carbon. Heat is given out when coal burns; this means merely that the chemical process of the oxidation of carbon into carbon dioxide is accompanied by the production of heat. In other words, the oxidation of carbon provides a supply of energy.

The body of every living animal is continually at work, and is therefore using up energy. This energy must somehow be supplied to it. In general, the supply of energy is obtained by combustion or, in other words, by the oxidation of carbon. This chemical process can occur, it need hardly be explained, without any visible flame; the combustion of a given quantity of carbon will provide the same amount of energy whether it takes place slowly, as in the human body, or rapidly and with the production of flame, as when coal is burned.

If energy is to be provided by the process of combustion, a supply of oxygen is necessary, whether the organism lives in water or out of it. The organism must, therefore, be provided with some means by which it can take oxygen from the atmosphere or from the water. In the lower forms of life the oxygen is absorbed through the skin and the carbonic acid or carbon dioxide, the end-product of combustion, is got rid of through the skin. In the course of evolution, special organs of breathing have been developed in many types of life; the gills of fishes, the respiring membranes of spiders and the lungs of human beings and many other animals are instances of such organs which, by increasing the area of the breathing surface, enable sufficient oxygen to be taken in to supply, through the process of combustion, the energy requirements.

Whenever, therefore, the supply of energy for the maintenance of vital processes is provided by combustion, an adequate supply of oxygen is necessary for the continuance of life. But is it possible that energy can be provided in any other way? There is one other source of energy, which is utilised by some organisms, viz. the process of fermentation.

The molecule of sugar, represented diagrammatically on p. 34, can be broken up into two molecules of alcohol:

$$\begin{array}{c} \text{H} \quad \text{H} \\ | \quad\ | \\ \text{H--C--C--OH} \\ | \quad\ | \\ \text{H} \quad \text{H} \end{array}$$

and two molecules of carbon dioxide. The chemical equation representing the transformation is as follows:

$$C_6H_{12}O_6 = 2\ C_2H_6O + 2\ CO_2$$

The sugar is broken up, alcohol being produced with evolution of carbon dioxide. This is merely the ordinary process of fermentation, which occurs, for instance, when wine is formed from grape-juice. The act of fermentation is accompanied by the evolution of heat, this heat representing the energy which can be made use of by any organism that relies upon fermentation for its supply of energy. The chemical change called fermentation, represented by the above equation, is brought about by yeast or other forms of cell life, which supply enzymes or transformers, whose presence makes possible a chemical change that would not otherwise take place.

The process of fermentation is not a very efficient method of providing energy; it is far inferior in efficiency to the process of combustion. It is consequently not well adapted as a general means of providing energy required for the maintenance of vital processes. It is the method employed, for instance, by certain parasitic organisms that live in the intestines, where a plentiful supply of starch or sugar is available. Such organisms do not need to provide energy for the maintenance of their own heat, because they live a sheltered existence under conditions of uniform temperature. They are able to live without any supply of oxygen, though they are dependent for their continued existence on a host who could not live without a supply of oxygen; they are consequently dependent in an indirect manner upon a supply of oxygen.

It seems reasonable to conclude that animal life is normally dependent upon a supply of oxygen, but that under exceptional circumstances it can exist without oxygen; it must then be able to obtain a supply of energy by the process of fermentation.

A further condition for the existence of life seems to be the presence of water, either in the liquid form or as water vapour. Neither seeds nor spores will germinate in absolutely dry soil; when life is not actually destroyed by the absence of moisture it becomes latent and no development ensues. Water is an essential constituent of the tissues of both animal and vegetable life, because a cell needs a certain amount of water to carry on its life. It is by imbibing water containing the chemical substances in which they feed that cells are enabled to grow and to multiply by continual division. In particular, we find that egg and sperm cells are specially sensitive cells, which cannot withstand much desiccation; in order that fertilisation may take place these cells must meet either in water or in a damp atmosphere such as is provided by the female sex ducts. In many animals that live in water, such as fishes and frogs, fertilisation takes place outside the body; in most animals that live on land, on the other hand, fertilisation takes place internally, within the female sex ducts. It does not seem probable that in the absence of moisture any form of life could develop.

There are many gases that have a marked toxic

action on living organisms, such as ammonia, chlorine, carbon monoxide and sulphuretted hydrogen. Though their presence in the atmosphere of any world might not necessarily prove conclusively that there could be no life on it, it would provide strong evidence against its probability. We shall find that poisonous gases are prominent constituents of the atmosphere of some of the planets.

The considerations summarised in this chapter should provide a reliable guide to the possibility or the impossibility of the existence of living organisms on another world, provided that we can obtain sufficient information about the conditions that prevail on it. If we can show that the conditions are favourable for life, it may not necessarily follow that there must be life. What is certain is that if suitable conditions exist, if there is an adequate supply of energy and if there is a suitable transformer for that energy, which can turn it into the chemical energy of carbon compounds, then the complex organic substances which form the basis of living cells not only can arise but will arise. How the step from these complex organic substances to the simple living cell is made is not known. Nevertheless, it seems reasonable to suppose that whenever in the Universe the proper conditions arise, life must inevitably come into existence. This is the view that is generally accepted by biologists.

It has been supposed by various people at one time or another that life did not arise on the Earth spontaneously, but that it was carried to the Earth by cosmic dust particles or by material particles

4

from some other world where life was already in existence. In some forms of this hypothesis, it is suggested that life was once created on one world and that it has ever since been disseminated from this original source. Such a theory does not bring us any nearer to understanding how life has come into being and merely removes the problem from the possibility of investigation by shifting it to some remote world. Helmholtz adopted a rather different form of the hypothesis and queried whether life had ever arisen; he suggested that life might be as old as matter and that the germs of life passed from one world to another and developed wherever they found suitable soil. All such hypotheses appear to be inherently improbable and they leave the solution of the problem of the origin of life as remote as ever.

If we accept the view that life on the Earth arose as the result of the operation of certain natural causes and that life will arise or has arisen on any other world where the proper conditions prevail, it does not follow that life will develop or that it has developed on such a world along the identical lines that it has followed on the Earth. The various forms of life that have appeared in succession on the Earth have no doubt resulted from the particular conditions, themselves slowly changing, that have prevailed on the Earth. If we consider the geological history of the Earth— the cooling down of the material in its initial gaseous condition until liquefaction took place, followed by the formation of a solid crust as further cooling ensued; the successive stages of crumbling and folding of the surface with the formation of mountain ranges, as the solid crust adjusted itself to a cooling and shrinking interior; the condensation of water vapour to form the oceans; the formation of sedimentary deposits beneath the oceans, through the denuding action of rain, streams and rivers, and their subsequent upheaval to form new land areas; the changes of climate shown by alternating ice-ages and glacial periods—it will be realised that life on the Earth, in the various forms that now exist, has been conditioned by a complex series of changes. The general sequence of events may have been similar on another world, but their details must inevitably have differed considerably. Differences in temperature, in the constitution of the atmosphere, in the proportions of land and water and in their distribution, and in the general accompanying phenomena must have profoundly influenced the course of evolution. The degree of adaptation to their environment shown by many animals provides evidence that mere environment has had a considerable effect on the development of life. It is reasonable to suppose, therefore, that life on any other world will have developed along forms that are entirely different from any with which we are familiar and that are possibly beyond our conception. Divergences, which may have been small in the most primitive types of life, would tend to become more and more accentuated in the slow course of evolution. Whether, on any other world where life may have arisen, intelligent beings —the counterpart of man—have evolved must be a question for mere surmise. It must remain outside the scope of this discussion.   The most that we are justified in attempting is a discussion of the suitability of the various planets of the solar system as a home of life and of the possibility that other planetary systems may exist amongst which there may also be some with conditions favourable for life.

CHAPTER III

METHODS OF INVESTIGATION

BEFORE considering what information can be obtained about the conditions that prevail on the various planets in the solar system, we must describe the methods by which some information can be gained about the extent and nature of their atmospheres and about the temperatures at their surfaces.

Simple considerations can tell us whether or not any planet may be expected to possess an atmosphere. Before looking at the planet in the telescope, we shall know what observation is likely to reveal.

Let us consider what an atmosphere really consists of. A gas is composed of an aggregation of molecules which are in a state of perpetual motion. The molecules move rapidly about in all directions, describing straight paths with uniform velocity, except when they collide with other molecules. If the gas is contained in a closed vessel, the continual bombardment of the walls of the vessel by the molecules that hit them produces an integrated effect which we call the pressure of the gas. If we double the quantity of gas in the vessel, the number of collisions of the molecules with the walls in each second is doubled; the pressure is therefore doubled. That is why the pressure in a motor tyre increases as we pump air into it.

If we think of a simple gas, all of whose molecules are of the same kind, at any given instant some molecules will be moving faster than others.

The speeds of some molecules will be reduced when a collision occurs; the speeds of other molecules will be increased. It is impossible to study in detail the motions of the individual molecules but the statistical distribution of the velocities amongst the complete aggregation of molecules can be investigated. The mathematical investigation of the statistical properties of the molecular assemblage forms the subject called the kinetic theory of gases. For a given temperature, the average velocity of all the molecules remains statistically constant. The number of fast-moving molecules with velocities greater than a value that is much in excess of the average velocity falls off very rapidly as this value increases; but there are definite, though small, proportions of molecules with speeds of 10, 20 or even 100 times the average speed.

The velocities of the molecules depend upon the temperature of the gas. The average velocity is proportional to the square root of the temperature measured from absolute zero ($-273°$ C.). At the absolute zero of temperature, the molecules have no velocity. They are all brought to rest. No lower temperature than this can be imagined. Physicists have been able to reach temperatures within a fraction of a degree of the absolute zero, but the zero itself has never been reached. The average velocity at the temperature of boiling water (100° C.) is, for instance, 17 per cent. greater than the average velocity at the freezing point of water (0° C.). By means of this relationship, the average velocity at any temperature can readily be found if its value at some given temperature is known.

In a mixture of gases, the molecules of different kinds move at different speeds. The lighter the molecules the faster they move on the average. There is a simple law, known as the law of equipartition of energy, which governs the distribution of velocities amongst the different kinds of molecules. The law of equipartition of energy states that in a mixture of gases the average energy of each kind of molecule is the same. Since the energy of a molecule is proportional to its molecular weight multiplied by the square of its velocity, it follows that the average speed of each kind of molecule is inversely proportional to the square root of its molecular weight. In a mixture of oxygen and hydrogen, for instance, the average speed of the molecules of oxygen will only be one-quarter of that of the molecules of hydrogen, because oxygen is sixteen times as heavy as hydrogen.

It is more convenient, in general, to use a velocity which is such that the pressure exerted by the gas is the same as if all the molecules were moving with this particular speed. It is slightly different from the average speed, being about nine per cent. larger. This velocity is tabulated for several gases, at the temperature of 0° C. (32° F.).

					Miles a second.
Hydrogen	.	.	.	.	1·15
Helium .	.	.	.	.	0·82
Water vapour .	.	.	.	.	0·38
Nitrogen	.	.	.	.	0·31
Oxygen .	.	.	.	.	0·29
Carbon dioxide	.	.	.	.	0·25

In a cubic inch of air at normal temperature and pressure (0° C. and 760 mm. of mercury) the number of molecules is about 500 million billion. The figures just given indicate that these molecules are moving with high speeds. It is evident that no molecule will move very far before it collides with another molecule. The average distance travelled by a molecule before it undergoes a collision is only about the two hundred thousandth part of an inch. The average distance traversed by a molecule between collisions is inversely proportional to the number of molecules in a unit volume and therefore becomes greater as the pressure is decreased. In a vacuum tube in which the pressure is one-tenth of a millimetre of mercury, the average distance between collisions is about one twenty-fifth of an inch. The average interval of time between successive collisions becomes correspondingly greater as the pressure is diminished.

The density of the gaseous atmosphere of the Earth or of another planet decreases with increasing distance from the surface of the planet. Near the upper limit of the atmosphere, where the density is very low, the molecules will travel for a considerable distance between collisions. If a molecule in this region happens to rebound after a collision in an outward direction and with a speed much greater than the average speed, there is the possibility that it may escape into outer space, provided that it does not come into collision with any other molecule.

In order that any particle, whether large or small, may be able to escape altogether in this way it is necessary that its velocity should exceed a certain critical value called the *velocity of escape*. The velocity of escape plays a very important rôle in our consideration of planetary atmospheres.

A gas possesses the property of spreading throughout the whole of any space in which it is placed. If, for instance, we place a small sealed flask containing gas in an evacuated chamber and then break the flask, the gas will at once spread throughout the whole of the chamber. This is the natural result of the movements of the molecules, which travel in straight paths until they collide with other molecules or with the walls of the containing vessel. Why, then, does not the atmosphere of the Earth rapidly dissipate away into space? Why do the outermost molecules not fly away? The reason is that the atmosphere is held bound by the gravitational pull of the Earth. The same force that makes the apple fall from the tree to the ground holds the air captive and prevents it spreading out into space.

Let us suppose that a stone is dropped from a certain height to the ground. It falls with an accelerated velocity because of the attracting force of gravity. We can imagine this motion to be reversed at each point of its path. If the stone is projected upwards with a velocity equal to that with which it hit the ground, its velocity will progressively decrease; at any distance above the ground, its upward velocity will be equal to its downward velocity at the corresponding distance when it was falling. The stone will come to rest instantaneously at a height equal to that from which it was previously dropped and it will then commence to fall to the ground again.

We now imagine the stone to be dropped from an infinitely great height, and we suppose the Earth to be isolated in space so that we need not concern ourselves with the gravitational attraction of any other bodies. The stone will fall towards the Earth with a gradually increasing velocity. It will reach the ground with a certain velocity, V, which will have a finite value, although the stone has fallen from an infinitely great distance. V is, in fact, given by the formula

$$V^2 = 2GM/a$$

where G is the value of the constant of gravitation, M is the mass of the Earth and a is its radius.

If the stone is projected upwards with the velocity V, it will reach an infinitely great distance before it comes to rest; if it is projected with a velocity less than V, it will eventually come to rest and then fall back to Earth, because any velocity less than V corresponds to the velocity acquired in falling to the ground from a height that is finite. It follows that the stone can only get completely away from the earth if its initial velocity is equal to, or greater than V. It is for this reason that V is called the velocity of escape.

We can determine the velocity necessary for any body to escape from the Earth by substituting the values for the constant of gravitation, and for the mass and radius of the Earth. In this way we find that the velocity of escape from the Earth is 7·1 miles a second.[1]

Returning to the consideration of the outer layers of the atmosphere, not a single molecule can possibly escape into outer space unless its velocity exceeds the escape velocity. But whenever a molecule rebounds away from the Earth with a speed greater than the escape velocity, it will escape from the Earth's gravitation, provided that it does not collide with any other molecule. There must inevitably be such a loss of the faster-moving molecules from the outer layers of the atmosphere.

Comparing the velocity of escape from the Earth with the average speeds of different types of molecules given above, it appears that a molecule of hydrogen must have rather more than six times its average speed to be able to escape, whereas a molecule of oxygen must have nearly twenty-five times its average speed. It is therefore much easier for hydrogen to escape than for oxygen.

---

[1] For those who are interested in verifying this value, the data are:

Constant of gravitation, $G = 6·67 \div 10^8$
Mass of Earth (in grams), $M = 5·97 \times 10^{27}$ gm.
Radius of Earth (in cms.), $a = 6·37 \times 10^8$ cm.
With these values we find
$V = 1·13 \times 10^6$ cm./sec. $= 11·3$ km./sec.
                                   $= 7·1$ miles a second.

But this statement is merely qualitative. Under any given conditions at what rate will the loss of atmosphere occur ?

This is a problem that is capable of treatment by the mathematical principles of the kinetic theory of gases. The necessary calculations were made some years ago by Sir James Jeans. He found that if the velocity of escape is four times the average molecular velocity, the atmosphere would be practically completely lost in fifty thousand years; if the velocity of escape is four and a half times the average molecular velocity, the atmosphere would be lost in thirty million years; whilst if the velocity of escape is five times the average molecular velocity, twenty-five thousand million years would be required for the loss to be almost complete.

The rate at which an atmosphere is lost is therefore conditioned in a very critical manner by the ratio between the velocity of escape and the average velocity of the molecules. If this ratio is 4, the rate of loss is very rapid, remembering that the age of the Earth is of the order of three or four thousand million years; if the ratio is 5, the rate of loss is so slow that we can regard the atmosphere as practically immune from loss. If, therefore, we know the velocity of escape from any planet (which is dependent upon a knowledge of the mass and radius of the planet) and the average velocity of the molecules (which is determined by the molecular weight and the temperature), we can estimate with considerable accuracy whether the planet is likely to have retained its original atmosphere almost in its entirety, or to have lost a substantial portion of its atmosphere, or to have lost essentially the whole of its atmosphere.

In the following table are given, for the Sun, Moon and planets, the radii and masses in terms of the corresponding quantities for the Earth as units, together with the velocities of escape in miles a second.

	Radius (Earth = 1).	Mass (Earth = 1).	Velocity of escape (miles/sec.).
Sun	109·1	332,100	392
Mercury . . .	0·39	0·044	2·4
Venus . . .	0·97	0·82	6·5
Earth . . .	1·00	1·00	7·1
Moon . . .	0·27	0·0123	1·5
Mars . . .	0·53	0·108	3·2
Jupiter . . .	10·95	317·1	38
Saturn . . .	9·02	94·9	23
Uranus . . .	4·00	14·65	14
Neptune . . .	3·92	17·16	15
Pluto . . .	? 0·10	? 0·01	? 2·2

Mere inspection of the figures for the velocities of escape, given in the last column, suggests that the large planets, Jupiter, Saturn, Uranus and Neptune, may be expected to have atmospheres that are much more extensive than the Earth's atmosphere; that Venus may be expected to have an atmosphere comparable with that of the Earth; that Mars may be expected to have an atmosphere considerably thinner than that of the Earth and that Mercury and the Moon may be expected to have little or no atmosphere. The extent to which these expectations are confirmed with be considered in subsequent chapters.

When a planet has been proved to have an atmosphere, we naturally wish to find out as much as possible about the composition of the atmosphere. For this purpose, we must have recourse to the spectroscope. The light from the planet is passed through the spectroscope, which contains one or more prisms; the result of this is to spread the light out into a band or spectrum, showing the colours of the rainbow, and each point in the spectrum corresponds to a definite wave-length.

If sunlight is analysed by the spectroscope, the light being admitted to the spectroscope through a narrow slit, it is found that the spectrum is crossed by a large number of fine dark lines, amounting to many thousands; to each of these lines there corresponds a definite wave-length and a definite intensity. They are known as the Fraunhofer lines, after the physicist who first investigated them. The analysis of this spectrum provides a great deal of information about the composition of the Sun. If, for instance, we pass an electric spark between two pieces of iron and examine the spectrum of the incandescent vapour between them, we find a considerable number of bright lines, spaced at irregular intervals and of different intensities; this particular series of lines is produced only by iron and by no other element. It is characteristic of iron; in a similar way, every other element has its own characteristic spectrum. If, now, light from a hot incandescent source, which has a continuous spectrum showing all the colours of the rainbow, is passed through the vapour of iron at a lower temperature, the continuous spectrum of the hot source is found to be crossed by a number of dark lines, each of which is exactly identical in wave-length with one of the bright lines in the spectrum of the incandescent iron vapour. Such a spectrum is called an absorption spectrum. The Fraunhofer lines in the spectrum of the Sun are of this nature; the hot interior of the Sun would give a continuous spectrum, but the cooler outer layers absorb the radiations of various wave-lengths, thus producing the dark Fraunhofer lines.

When we investigate the spectrum of sunlight, we find that the spectrum of iron, line by line, is contained in it, proving conclusively that there is iron in the Sun. In a similar way, we can detect the presence of one element after another in the Sun and learn much about the elements contained in the Sun ; we can, moreover, go farther than this and use the relative intensities of the lines due to different elements to obtain some fairly reliable conclusions about the abundance of each element. If, for instance, we were to double the amount of one element in the Sun, leaving the amounts of the other elements unaltered, we should find the intensities of the lines of that particular element would be relatively strengthened. It is by means of such considerations that we can determine the relative abundance of this or of that element in the Sun, or in a remote star or nebula.

There is, however, one complicating factor that enters into the analysis of the light from the Sun. We make our observations from the bottom of our extensive atmosphere. The light from the Sun has to pass through this atmosphere before it reaches us and some of the light is absorbed in the atmosphere.

The consequence is that some of the absorption lines that are present in the observed spectrum of the Sun do not originate in the Sun but in the atmosphere of our Earth. Of particular importance to us is the absorption produced by the ozone in the atmosphere. The amount of ozone in the atmosphere is extremely small; it is estimated to be equivalent to a layer about one-tenth of an inch thick at atmospheric pressure and room temperature. It occurs almost exclusively above the highest clouds, the greatest density being at a height of between twenty and thirty miles. Small though the amount of ozone is, the absorption produced by it in the ultra-violet region of the spectrum is so strong that all of the light of wave-length shorter than 0·000012 inch is completely absorbed; none of the light in this region of the spectrum is consequently accessible to observation.[1] Unfortunate though this is for the investigations of the astronomer, it is a fortunate circumstance for life as we know it: for animals, including human beings, could not exist as they are now constituted, if there were not a small amount of ozone in the atmosphere.

The importance of ozone biologically is this:

[1] The approximate limits of wave-length for the different colours of the spectrum are as follows:

Ultra-violet	∠ 0·000014 inch.
Violet	0·000014 to 18 inch.
Blue	0·000018 to 20 inch.
Green	0·000020 to 22 inch.
Yellow	0·000022 to 24 inch.
Orange	0·000024 to 26 inch.
Red	0·000026 to 30 inch.
Infra-red	⊐ 0·000030 inch.

much of the radiation of short wave-length that is absorbed by ozone is very injurious to the eye and has an injurious effect also on the other tissues of the body. The light from glowing mercury vapour is rich in the ultra-violet light. Quartz is fairly transparent to such light, whilst glass absorbs it strongly. A quartz mercury vapour lamp is therefore a powerful source of ultra-violet light and such lamps are accordingly used for ultra-violet light treatment; anyone who has undergone such treatment knows that the exposure to the lamp must be carefully timed, otherwise severe damage to the tissues may result, and that it is necessary to wear dark spectacles for the protection of the eyes during the exposure to the rays. In moderate amount, however, the ultra-violet light is beneficial because it keeps us in proper health and is effective in pre-venting rickets. If, then, there were no ozone in the atmosphere, eyes as now constituted could not have developed and the bodily tissues would be seriously injured; if, on the other hand, the ozone were present in greater amount, life, as we are familiar with it, could not continue. It is not to be concluded that no animal life would be possible if there were no ozone in the atmosphere; this is merely a striking illustration of how life has adapted itself to the conditions that prevail. In this particular instance, however, the adaptation is surprisingly close. The amount of ozone in the atmosphere is not constant, it is variable within somewhat narrow limits but these limits fortunately lie between the two extremes, the one at which on the one hand the amount of ultra-violet light would become destructive of life and the other at which the amount would become insufficient to maintain it.

The oxygen in the atmosphere produces some strong absorptions in the spectrum of the Sun; these are mainly in the regions of longer wave-length, the red and the near infra-red regions of the spectrum, though there are also some weaker absorptions of shorter wave-length in the visual spectrum. The two strong absorptions in the red region of the spectrum that were called the A and B bands by Fraunhofer are caused by oxygen. Water-vapour in the atmosphere produces some extremely strong absorptions in the long wave-length infra-red region. Nitrogen, on the other hand, though it is the most abundant constituent of the atmosphere, produces no absorptions in the region with which our investigations are concerned.

When we analyse the spectrum of the Sun, how are we to distinguish between the absorptions that originate in the Sun itself and those that are produced by our atmosphere? There are two ways in which the absorptions of terrestrial origin can be identified. The first method is to compare the spectra of the Sun taken at different altitudes. The lower the altitude of the Sun, the longer is the path that the light from the Sun has to travel through the atmosphere in order to reach us; it is, in fact, for this very reason that the light from the Sun is much less intense towards sunset than it is at midday. The longer the path through the atmosphere the greater become the effects of atmospheric absorption. The lower the altitude of the Sun, therefore, the more intense are the absorptions of terrestrial origin relative to those of solar origin. The second method is to compare the spectra of light from the east and west limbs of the Sun. The Sun rotates on its axis in about twenty-seven days; as a result of this rotation the west limb is moving away from us whilst the east limb is moving towards us. The wave-length of the light received from a source that is moving towards us is shortened and the absorptions are therefore displaced slightly towards the violet end of the spectrum ; the wave-length of the light from a source that is moving away from us is lengthened and the absorptions are therefore displaced slightly towards the red end of the spectrum. Comparing the spectra of the east and west limbs of the Sun, we consequently find that there is a slight separation between the corresponding absorptions in the light from the two regions. The absorptions that originate in our atmosphere are not affected by the rotation of the Sun and these absorptions coincide in position in the spectra of the light from the two opposite limbs. If, then, the spectra of the eastern and western limbs of the Sun are photographed simultaneously or in immediate succession on the same plate, the absorptions of terrestrial origin can be at once picked out, because they consist of all the absorptions that coincide in position in the two spectra. This method has the advantage over the first method that it does not necessitate the comparison of spectra obtained at different altitudes and therefore at different times.

Having by one or other of these methods identified the absorptions that originate in the atmosphere of the Earth, the way is open to learn something about the composition of the atmospheres of the planets. Let us first consider for a moment the nature of the light that we receive from a planet. The planets are cool bodies and have no intrinsic light of their own. We see a planet by means of light from the Sun that falls upon it and is reflected back. As the sunlight penetrates into the atmosphere of the planet, it is partially scattered and partially absorbed. The depth to which it penetrates depends upon the nature and extent of the atmosphere; the light may or may not actually penetrate to the surface of the planet. Some of the light that reaches us from the planet will have penetrated to a greater depth into the atmosphere of the planet and some will have penetrated into a lesser depth; the net effect may be expected to be that the light will bear some impress of its passage into and out of the atmosphere of the planet, so that when the light is analysed by the spectroscope, absorptions that have originated in the atmosphere of the planet may be revealed, which will serve to give some clues to the nature of the atmosphere. An absorption in the atmosphere of the planet that does not correspond with any absorptions produced in the atmosphere of the Earth will be readily revealed. On the other hand, it may happen that the atmosphere of the planet contains oxygen or water-vapour, whose absorptions will coincide with the absorptions produced by the same substances in the Earth's atmosphere. Some care is required, therefore, to decide whether the absorptions produced by oxygen and water-vapour arise entirely in the atmosphere of the Earth or whether they include the effect of absorptions originating in the atmosphere of the planet. Oxygen and water-vapour are the two substances whose presence in the planetary atmosphere is of the greatest significance for the possibility of the existence of life and it is these two substances whose presence may be most difficult to establish.

Two methods can be used to distinguish between an absorption of terrestrial origin and the same absorption of planetary origin. One method is to compare the spectrum of the planet with the spectrum of the Moon, the two spectra being obtained on the same night and as nearly as possible at the same time. The Moon and the planet should be at equal altitudes when their spectra are photographed, so that the absorptions produced by the Earth's atmosphere will be equal, or very nearly equal, in the two cases, because the air-paths are equal. The Moon, as we shall see, is devoid of atmosphere; it follows that if an absorption is present in the spectrum of the planet and not in that of the Moon, or is stronger in the spectrum of the planet than in that of the Moon, it must originate in the atmosphere of the planet. If the absorption produced by the atmosphere of the Earth is very much more intense than that produced by the atmosphere of the planet, we may fail to detect in this way the planetary absorption, because the difference in the intensities of the absorptions in the two spectra may be too small. The second method of investigation is a more delicate one, and is specially useful for deciding whether substances that are present in our own atmosphere are also present in the atmosphere of the planet. It is based on the displacement of the absorption lines in the light from a moving source, which we have already used for the detection of the absorptions produced by the Earth's atmosphere from the comparison of the light from the east and west limbs of the Sun. The spectrum of the planet is photographed at a time when the planet is approaching or receding from the Earth most rapidly. The relative motion will displace the absorptions due to the planet's atmosphere with respect to those due to our own atmosphere and in this way we may hope to detect planetary absorptions of low intensity.

It must be emphasised that though by these methods we may expect to obtain some information about the constitution of the atmosphere of the planets, the information can never be complete. There are many possible constituents of an atmosphere that show no absorptions in the region of the spectrum that is accessible to study. There is no means available by which such constituents can be detected. Amongst these undetectable constituents are hydrogen, nitrogen, helium, neon and argon.

A knowledge of the temperatures of the planets is of importance for our consideration of the possibility of the existence of life. Some general information about the temperature conditions can be obtained from theoretical considerations.

We shall assume, to begin with, that the planet has no atmosphere, because the effects of an atmosphere on the temperature of the planet are complicated and difficult to estimate. We shall consider first the extreme case of a planet which, like Mercury, always turns the same face to the Sun. We can estimate the temperature of the surface of the planet if we suppose that the planet has no output of heat of its own, but that there is an exact balance between the heat that it receives from the Sun and the heat that it radiates into space. From the measurements of the radiation from the Sun received at the surface of the Earth, we know that the energy from the Sun reaching the Earth just outside the Earth's atmosphere is equivalent to 1·54 horse-power each square yard. If the distance of the planet from the Sun is R times the distance of the Earth from the Sun, the point on the surface of the planet where the Sun is overhead receives energy at the rate of $1.54/R^2$ horse-power per square yard. This must be equivalent to the rate at which heat is radiated. This rate is determined by the temperature of the planet. In 1879 Stefan put forward the suggestion that the total radiation from a body is proportioned to the fourth power of its absolute temperature (i.e. the temperature measured from the absolute zero, — 273° C.). This law is accurately true only for what is termed a " black body," a body that completely absorbs all the radiations falling on it. The radiation from the planets is not strictly in accordance with Stefan's law, though we can use the law to derive an estimate of the temperature that will be sufficiently accurate for our purpose. The constant of proportionality, which enters into Stefan's law, has been found by experiment. Using this observed value and equating the energy received and radiated, it is found that the temperature (on the absolute scale, denoted by K) is given by the expression $T = 392°/\sqrt{R}$. At the distance of the Earth, where $R = 1$, the temperature obtained is 392° K, or 119° C. The temperature found in this way is the highest temperature at any point on the surface of a planet that always turns the same face to the Sun—provided the assumptions that the planet has no atmosphere and that it is a "black body" are satisfied. At other points on the planet the temperature would be lower, because the Sun's rays would fall obliquely instead of vertically; the dark side of the planet, which receives no heat from the Sun, would be extremely cold.

If the planet does not always turn the same face to the Sun, the effect of its rotation would be to lower the noonday temperature and to raise the night temperature; the noonday temperature would be lowered because no point of the surface would have the Sun in its zenith for more than a short time and the night temperature would be higher because the cooling off of the surface that had been heated by the Sun in the daytime would be gradual. The faster the rotation, the smaller would be the difference between the day and night temperatures. For a sufficiently fast rotation, there would be no difference between day and night temperatures at any place on the planet; there would be, however, a variation of temperature with latitude, because the average rate of reception of heat from the Sun depends on the latitude. The average temperature over the whole surface under these conditions can be estimated.

Suppose the planet to be at the distance of the Earth and its radius to be r yards. The planet is receiving energy from the Sun at the rate of $1\cdot54 \times \pi r^2$ horse-power, because the cross-sectional area exposed to the radiation is πr^2. If the distance of the planet from the Sun is R times the distance of the Earth from the Sun, the corresponding rate is $1\cdot54 \times \pi r^2/R^2$ horse-power. The total radiation of energy from the whole surface ($4\pi r^2$) is given, in terms of the temperature T, by Stefan's law. Equating the energy received with the energy radiated, we obtain $T = 277°/\sqrt{R}$, expressed on the absolute scale. It may be noted that the average absolute temperature under these conditions will be equal to the maximum temperature previously found, divided by $\sqrt{2}$. The average temperature for the Earth under these conditions would be 277° K or 4° C.

The temperature on a planet that possesses an atmosphere cannot be so readily calculated, because complex meteorological effects come into play and, as is well known on the Earth, temperatures at any one place may differ considerably from day to day. The general effect of an atmosphere is nevertheless readily seen to be a smoothing out of the temperature differences between day and night, because there will be a persistent tendency for heat to be carried from the warmer to the cooler parts of the surface by warm air moving into colder regions and cold air moving into warmer regions.

The temperature at the surface of a planet is influenced by its atmosphere in another way. The atmosphere of the Earth, and probably also most planetary atmospheres, are opaque in many regions of the infra-red, corresponding to long wave-length radiations. Most of the solar heat is transmitted by the atmosphere, warming the surface of the planet; much of this heat is radiated again as radiation of considerably longer wave-length, to which the atmosphere is opaque. The temperature is therefore raised considerably. By preventing the escape of the radiations of long wave-length, the atmosphere exerts a blanketing effect and the fall of temperature at night becomes less rapid; the diurnal range of temperature is therefore appreciably reduced.

The table on p. 76 gives a comparison between the measured temperatures and the temperatures estimated on the two hypotheses mentioned above. The second column gives the directly measured temperatures; the third column gives the average temperatures on the assumption that there is no diurnal variation of temperature; the fourth column gives the maximum temperature on the sunlit face, assuming that the same face is always turned to the Sun. The planets are arranged in the table in the order of increasing distance from the Sun, and therefore also of decreasing temperature. If the distance of the Earth from the Sun is represented by one foot then, on the same scale, the distances of the other planets are approximately as follows: Mercury, 4½ inches; Venus, 8½ inches; Mars, 18 inches; Jupiter,

5 feet; Saturn, 9½ feet; Uranus, 19 feet; Neptune, 30 feet; Pluto, 40 feet.

The radiation received on the Earth from the planets can be measured with the aid of a large telescope, to gather as much of the radiation as possible, in conjunction with a sensitive detector of radiation. For detecting and measuring the radiation a bolometer or a thermocouple may be used. In the bolometer the radiant energy is focused on to a minute strip of platinum, which forms one arm of an electrical circuit known as a Wheatstone's bridge. A similar strip, shielded from the radiation, forms a second arm of the bridge, which is balanced against the first. When the radiant energy falls on the first arm of the bridge, it is heated, its resistance is increased, the balance of the bridge is upset and a current flows through the galvanometer of the bridge. The deflection of the galvanometer provides a measure of the intensity of the radiation falling on the bolometer. The thermocouple consists of a small junction of two minute strips of different metals; when the junction is heated a thermoelectric current, whose strength is proportional to the intensity of the radiation, flows through the circuit and is measured by a sensitive galvanometer. A highly sensitive thermocouple will detect the heat from a candle at a distance of three miles.

The thermocouple or bolometer provides a measure of the total radiation from the planet, as modified by absorption in the Earth's atmosphere. A correction must be applied to the measures to allow for this effect, but we need not enter into the

details of this correction. When it is applied we
obtain a measure of the total planetary radiation,
which consists in part of reflected sunlight and in
part of the low-temperature long-wave radiation
from the planet itself. It is the latter portion that
provides information about the planet's temperature and it is necessary, therefore, to separate it from
the portion that is merely reflected sunlight. The
separation is easily effected by placing a small
transparent vessel containing water in the path of
the rays. The water transmits the portion of the
radiation of relatively short wave-length—the
reflected sunlight portion—but is opaque to the
long-wave planetary portion. We are thus enabled
to measure the true heat radiation from the planet.
The measured temperatures are given in the second
column of the table.

PLANETARY TEMPERATURES

	Measured	Calculated	
	° C.	I ° C.	II ° C.
Mercury (mean, sunlit side) .	400	172	358
Venus (bright side) . .	55	54	191
Venus (dark side) . .	— 20	—	—
Earth	14	4	119
Moon (centre of sunlit side) .	120	4	119
Moon (centre of dark side) .	— 150		
Mars (hottest portions) . .	20	— 51	43
Jupiter (average) . .	— 140	— 151	— 100
Saturn (average) . . .	— 155	— 183	— 145
Uranus (average) . . .	< — 180	— 210	— 184
Neptune	—	— 222	— 201
Pluto	—	— 229	— 211

It will be noted that in general the measured temperatures lie between the temperatures given in columns three and four, and that there is a close general agreement between the estimated temperatures and those that are actually observed. The measured temperatures will be commented upon in subsequent chapters when dealing with the individual planets.

CHAPTER IV

THE EVOLUTION OF THE ATMOSPHERE OF THE EARTH

IN the last chapter we found that the loss of atmosphere from a planet will be very rapid if the velocity of escape from the planet is less than four times the average velocity of the molecules and that the atmosphere will be practically immune from loss if the velocity of escape is more than five times the average velocity of the molecules. Let us apply these conclusions to the Earth and see whether the atmosphere of the Earth is in accordance with expectations.

The velocity of escape from the Earth is 7·1 miles a second. Any component of the atmosphere, whose average molecular velocity is less than about 1·4 miles a second, should therefore be practically immune from loss. The average molecular velocities for different gases are given on p. 56; it will be seen that they are all less than the critical value of 1·4 miles a second. These velocities correspond to a temperature of 0° C.; for higher temperatures the velocities will be greater. The temperatures that occur on the Earth at the present time are not great enough, however, to bring the average molecular velocity of hydrogen to a value as great as 1·4 miles a second; this velocity requires a temperature of 88° C.

It appears, then, that the atmosphere of the Earth should be immune from the loss of hydrogen at the present time and, therefore, immune also from the loss of all other gases. The Earth should, therefore, have retained the whole of its initial atmosphere, unless the rate of loss of the atmosphere soon after the Earth was formed, during the period when the Earth was very much hotter than it now is, was so rapid that in the course of a few thousand years much of the atmosphere was lost. There is evidence, which we shall consider presently, that such a loss did indeed occur.

The estimated composition of the atmosphere, by volume and by weight, according to the figures given by Humphreys, is as follows:

COMPOSITION OF EARTH'S ATMOSPHERE

Constituent	Volume, per cent. of dry air, at surface.	Weight in units of 100 million tons.
Total atmosphere . .	—	50,293,000
Dry air	100·00	50,162,360
Nitrogen	78·03	38,111,360
Oxygen	20·99	11,413,080
Argon	0·9323	609,040
Water-vapour . . .	—	130,490
Carbon dioxide . .	0·03	21,316
Hydrogen . . .	0·01	1,270
Neon	0·0018	678
Krypton	0·0001	126
Helium	0·0005	79
Ozone	0·00006	29
Xenon	0·000009	17

In addition, there are varying quantities of impurities such as sulphur compounds, ammonia, nitric and nitrous acids, particles of salt from the

PLATE 6

THE MOON: REGION OF COPERNICUS

Copernicus is the large crater at the centre of the photograph. It has a diameter of 56 miles. The mountain-ring, which reaches a height of 11,000 feet above the floor of the crater, falls gradually on the outside but very precipitously, with deep ravines, on the inside. The white streaks or rays, which radiate from Copernicus, are well shown; it will be noticed that they pass over mountains and across craters.

To the right of Copernicus are some hundreds of small craters, about 400 to 500 feet in diameter, many of which are arranged in rows.

The range of mountains in the upper left-hand portion of the photograph is known as the Carpathians.

Photographed by Dr. J. H. Moore and Mr. J. F. Chappell, with the 36-inch refractor, Lick Observatory, 1937, October 26.

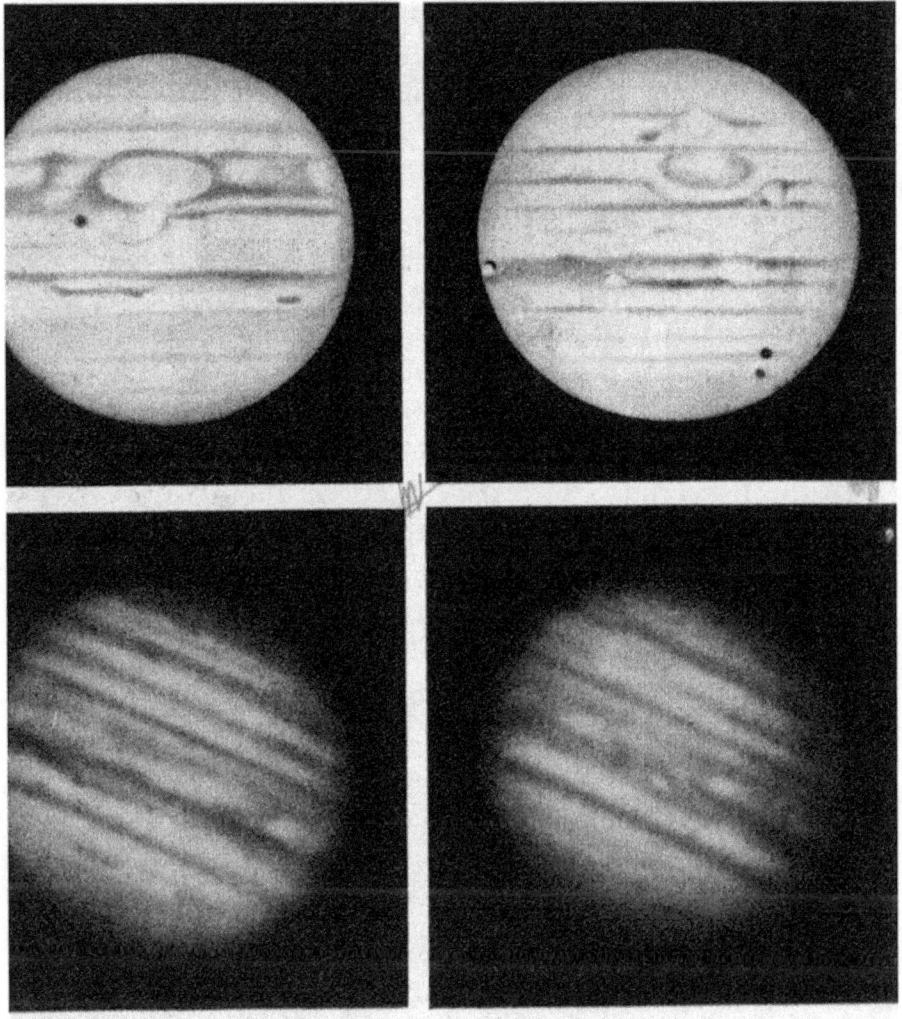

PLATE 7

THE PLANET JUPITER

The upper portion of the plate shows two drawings of Jupiter by Rev. T. E. R. Phillips. The left-hand drawing was made on 1908, April 24, and shows Satellite III (Ganymede) as a dark spot in transit across Jupiter. The drawing shows the common appearance of the Red Spot, seen as a prominent oval near the satellite, during conjunction with the South Tropical Disturbance. The right-hand drawing was made on 1933, March 9. It shows Satellite I (Io) partly occulting its shadow, near the left-hand limb of the planet's disk. Satellite IV (Callisto) is shown in transit on the lower part of the disk and casting its shadow on the disk.

These two drawings show nearly the same presentation of the planet and illustrate the changes in the markings, which are atmospheric markings and not surface details.

In the lower portion of the plate two photographs of Jupiter taken by Dr. Jeffers with the 36-inch refractor of the Lick Observatory on 1939, October 13 (left) and October 21 (right). These photographs show clearly the complex system of dark belts on Jupiter. In the right-hand photograph the Red Spot may be seen above and to the left of the centre.

evaporation of sea spray, particles of soot, fine dust, and pollen of many varieties.

The existence of the rare gas, argon, in the atmosphere was discovered by Lord Rayleigh and Sir William Ramsay in the year 1894. It may appear surprising that, although one per cent. by volume of the air we breathe is argon, this gas was not discovered at a much earlier date. The reason is that argon is a very inert gas and does not form chemical compounds, so that its existence was not recognised. It will be seen from the above table that the weight of argon in the atmosphere is several times the total weight of the water-vapour, carbon dioxide, and every other atmospheric constituent, except the two principal constituents, nitrogen and oxygen. In the years 1895–98, shortly after the discovery of argon, Sir William Ramsay and his assistant, Travers, discovered the four other rare gases, helium, neon, krypton and xenon.

Helium, as previously mentioned, is the only substance that was discovered in the Sun before it was found on the Earth. First discovered on the Earth by Ramsay as a constituent of the atmosphere, it was later found to be present in cleveite and other uranium- and thorium-bearing minerals; helium gas, given off by such minerals, is collected from certain bore-holes in the United States. Helium being the lightest gas with the exception of hydrogen, and not forming an inflammable mixture with air, has been used for filling the balloons of airships.

Argon is obtained today in a state of comparative purity from commercial oxygen-distilling apparatus and some ten million cubic feet are used each year for filling electric-light bulbs, the so-called gas-filled bulbs, because the efficiency of the lamp is thereby increased and its life is prolonged.   Neon is extensively used for advertising signs.   Krypton and xenon have not been used much for commercial purposes but they may eventually be used in electric-light bulbs, as they increase the efficiency of the lamp still further.

The amount of water-vapour in the atmosphere is very variable, depending upon temperature and other conditions.   Near the Earth's surface, the proportion of water-vapour by volume may vary from a mere trace to about five per cent. on hot days of very high humidity.   The water-vapour is found only in the lowest layers of the atmosphere; there is very little above a height of five miles, because of the low temperature at the higher levels. If all the water-vapour in the Earth's atmosphere were condensed, the condensed water would be sufficient in amount to cover the whole of the surface of the Earth to a depth of about one inch.

The composition of the atmosphere in its lower layers, up to a height of about ten miles, is practically constant, except for the water-vapour content; in this region there is continual mixing of the constituents by convection and the action of turbulence.   In the upper regions of the atmosphere there is very little vertical movement and the distribution of the constituents is therefore controlled by the action of gravity.   The percentage of the lighter gases must, therefore, increase in the upper regions, and at the highest levels hydrogen and helium must be the principal constituents.

It is not possible to state how high the atmosphere extends above the surface of the Earth. The density of the atmosphere decreases with height, gradually thinning out until empty space is reached; but there is no sudden transition from air to empty space. Shooting stars become visible at heights of from 70 to 100 miles; these small bodies, which enter the Earth's atmosphere from outside, are not seen until the friction caused by their rapid motion through the air heats them to such an extent that they become incandescent. The aurora borealis, or northern lights, gives evidence of the extension of the atmosphere to much greater heights. The auroral light is an electrical phenomenon, caused by the entrance of electrified particles into the atmosphere. By taking photographs of the aurora from two places, some miles apart, the height can be found. The lower edge of the aurora is usually at a height of some 60 or 70 miles; the highest portions have been found to extend to heights as great as 500 to 600 miles. The density at such heights is so extremely small that we may regard the effective limit of the atmosphere as being about 600 miles above the surface. If the atmosphere were compressed so that it was of uniform density throughout, the density being equal to the actual density at the surface, it would extend to a height of only $5\frac{1}{2}$ miles; this is called the height of the *equivalent atmosphere*.

It is of some interest to note that the total weight of the Earth's atmosphere is rather less than a one-millionth part of that of the Earth itself and that the weight of the atmosphere is equivalent to that of an ocean of water covering the whole of the Earth's surface to a depth of about thirty-three feet. The least abundant of the constituents, xenon, if loaded on railway wagons, each carrying ten tons, would require a train to carry it that would extend eighty times around the Earth's equator. Travelling at twenty miles an hour, such a train would take twelve years to pass by.

We have seen that helium is present in the atmosphere to the extent of about five parts in one million by volume. Helium is being continually added to the atmosphere by the process of the weathering of the igneous rocks of the Earth's crust, which contain uranium and thorium. In any mineral or rock that contains these elements, radioactive disintegrations are continually taking place, one result of the break-up of these heavy atoms being the formation of helium. Some of the helium so produced remains inside the minerals and rocks and some escapes into the atmosphere; the proportion that escapes depends on various geological factors. When such rocks are decomposed by weathering, the whole of their helium escapes into the atmosphere. It has been estimated that the atmosphere does not now contain more than a fraction of the amount of the helium that it has gained during geological times in the process of the formation of the sedimentary rocks as a result of the weathering of the igneous rocks. It must be concluded that much of the helium that has been added to the atmosphere in this way has

somehow escaped; the suggestion has been made
that there may be at the present time an approximate balance between the amount of helium that
is still being added in the way we have described
and the amount that is being lost.

It may be objected that the loss is, possibly, only
apparent and that the helium, being a light gas,
has become concentrated in the upper regions of
the atmosphere where little or no mixing up of the
constituents can occur by convection. There is
direct evidence, however, that the upper atmosphere cannot be rich in helium because the
spectrum of light from the aurora, which comes
from a height of something like sixty miles, shows
the presence of oxygen and nitrogen but not that of
helium.

Even if the Earth had remained sufficiently hot
in the early stages of its existence for a time long
enough for the hydrogen and helium then present
in its atmosphere to escape entirely, it still remains
to explain how helium continues to be lost when,
according to the theoretical conclusions already
mentioned, which are based on the accepted principles of the kinetic theory of gases, it should not be
possible for it to escape. There is one process by
which the escape of helium can be brought about.
It is well known that the night sky is faintly luminous. In addition to the light from the stars there is
a faint luminescence from the upper atmosphere,
whose brightness seems to vary with the sun-spot
cycle, being greater at sun-spot maximum than at
sun-spot minimum. Lord Rayleigh has termed
this luminescence the non-polar aurora. In the

spectrum of this faint light from the night sky, obtained by taking long exposures with special spectographs of small dispersion, the characteristic green and red lines that occur in the spectrum of the bright aurora borealis are always present. These particular lines are known to be produced by radiations from atoms of oxygen that are in a special condition, which the physicists term a *metastable* state.   An atom, when excited or loaded up with energy, usually unloads its energy within a short interval of time of the order of one hundred-millionth of a second; this unloading of the energy corresponds to the emission of radiation.   A meta-stable state, on the other hand, is characterised by the peculiarity that the atoms in that state do not have a very strong inclination to unload their energy: they may remain for an average time of a second or longer before emitting their energy in the form of radiation.   But collisions between the atoms are so frequent that there is a high proba-bility that before an atom in the metastable state emits its energy as radiation it will have collided with another atom.   Whenever a collision of a metastable oxygen atom with another atom occurs, the energy of the oxygen atom—which in due course would have been emitted as radiation—is immediately unloaded and is converted into kinetic energy.   Instead of being emitted as radiation, it is used in making the two colliding atoms rebound with a greatly increased speed.   The amount of energy that is unloaded in this way when another atom collides with a metastable oxygen atom can be computed.   It is found that if the colliding atom is an atom of helium the energy is sufficient to enable the atom of helium to rebound with a speed of more than seven and a half miles a second.   As this speed is greater than the velocity of escape from the Earth, the atom of helium has a chance of escaping, which it would not otherwise have. Hydrogen atoms would acquire a still greater speed and could also escape.   Heavier atoms, such as those of nitrogen and oxygen, though they would receive an equal amount of energy by the collisions, would not acquire such large velocities; they would not rebound with a speed greater than the velocity of escape.   The loss of hydrogen and helium from the atmosphere of the Earth at the present time is thus made possible by this special process, depending upon the fact that free oxygen is present in the atmosphere.   It is possible in this way to explain satisfactorily why the amount of helium in the atmosphere at the present time is less than we should have expected.

There is strong evidence that the present atmosphere of the Earth is not its original atmosphere and that the primitive Earth must have remained hot sufficiently long for most of the initial atmosphere to have been lost.   This evidence is provided by the comparison between the relative abundance of different elements on the Earth and the relative abundance of the same elements in the Sun and the stars.   We must here anticipate a subject that will be discussed more fully in a later chapter.   It is believed that the Earth, in common with the other planets in the solar system, was formed from matter drawn out from the Sun by the gravitational action of another star that passed close by the Sun. If this was so, and there is no satisfactory alternative theory to account for the origin of the solar system, we should expect that the composition of the Earth would be generally similar to that of the outer layers of the Sun.

The investigation of the chemical composition of the atmosphere of the Sun is based upon the study of its spectrum; the number of atoms needed to produce a line of given intensity in the spectrum of the Sun can be estimated. The composition of the Earth is estimated from the actual analysis of typical rock samples, the results being combined in proportion to the relative amounts of the different kinds of rocks, as inferred from geological evidence. This information is supplemented by information about the interior of the Earth provided by the study of seismograph records of earthquakes, which provide information about the propagation of the earthquake waves through the interior of the Earth, and hence about the nature of the interior. There are, naturally, uncertainties attaching to both methods, because in neither case is the available data as complete as we should like. The comparison of the conclusions from the two entirely different methods of investigation is, nevertheless, of great interest. The most reliable information about the composition of the atmosphere of the Sun is obtained for the metallic elements. Russell gives the following table, in which the fourteen most abundant metals in the Sun, found from his investigation of the Sun's spectrum, are compared with the fourteen most abundant metals in the Earth, as indicated by the work of Gold-schmidt:

COMPARISON OF ABUNDANCE OF ELEMENTS IN EARTH AND SUN

Group.	Earth.	Sun.
I . . .	Iron	Magnesium
	Magnesium	Sodium
	Aluminium	Iron
	Nickel	Potassium
	Calcium	Calcium
	Sodium	Aluminium
	Potassium	
II . . .	Titanium	Manganese
	Chromium	Nickel
	Manganese	Chromium
	Cobalt	Cobalt
		Titanium
III . . .	Copper	Vanadium
	Vanadium	Copper
	Zinc	Zinc

In the above table, the elements are divided into three groups. The average abundance of elements in the first group is about ten times that of elements in the second, and the average abundance of elements in the second group is about ten times that of elements in the third.

The same fourteen metals appear in the two columns headed Earth and Sun and the grouping into the three groups is generally similar.

The analysis of meteorites gives the following as the most abundant metallic elements: iron, magnesium, sodium, nickel, calcium, aluminium. The meteorites are probably fragments of matter drawn from the Sun when the planets were formed, which did not condense into planets. It will be noticed that the six most abundant metals in the meteorites are included in Group I in the list of most abundant metals in the Earth.

The investigation has been extended to include the metals of lower abundance than those given in the above table; when the lists of relative abundance for the Earth and the Sun are compared, it is found that out of a total of forty-eight metals there are only four cases in which one list makes a given element more than ten times as abundant as the other, though some of these metals are a hundred thousand times more plentiful than others. The four cases of greatest discordance all relate to metals for which either the spectroscopic data for the Sun or the chemical data for the Earth are known to be rather uncertain. The conclusion drawn by Russell from this detailed investigation is that " it is hard, indeed, to find a single case in which we can be sure that a given metal is more or less abundant in the Sun than on the Earth."

This striking similarity in composition may be regarded as providing substantial evidence of a common origin. It is just what we might have anticipated if the Earth was formed from material that had been drawn out from the Sun. It is rather more surprising to find that a detailed investigation of the spectra of the stars, made in a manner analogous to that used for the analysis of the spectrum of the Sun, shows that the composition of the stars is substantially similar to that of the Sun. This seems to suggest that the Sun and all the stars have themselves been formed from some primæval material whose composition was everywhere more

or less the same and that subsequent processes of
building up more complex atoms out of simpler
ones, or of the breaking up of the most complex
atoms, have followed a generally similar course.

When we turn from the metals to the non-metals
we find an entirely different picture.   The striking
similarity between the composition of the Earth
and that of the Sun and the stars no longer persists.
Some remarkable differences are shown.   Hydrogen, for instance, is far more abundant in the Sun
than in the Earth.   At least 90 per cent. of all the
atoms in the atmosphere of the Sun, and perhaps
95 per cent. or more, are atoms of hydrogen ;
there are some three hundred times as many atoms
of hydrogen in the Sun as there are atoms of all
the metals together.   The number of atoms of
hydrogen in terrestrial rocks is about equal to the
average number of atoms of the six most abundant
metals contained in the rocks : aluminium, iron,
calcium, sodium, potassium and magnesium.   It
might be thought that the oceans would account
for much of the terrestrial hydrogen, as each molecule of water contains two atoms of hydrogen; but
the oceans only contribute 239 parts by weight out
of one million parts for the Earth as a whole.

Another striking discordance is provided by
nitrogen.   Nitrogen is very abundant in the Sun,
the stars and the nebulæ.   The nitrogen in the
Earth's atmosphere amounts to about one part by
weight in two million for the Earth as a whole;
there is a little nitrogen in the igneous rocks and
some may be dissolved in the liquid core of the
Earth.   But with ample allowance for such possi-

bilities, nitrogen must be at least some hundreds of times more abundant in the Sun and stars than in the Earth.

Perhaps the most interesting discordances are those shown by the rare gases of the atmosphere, argon, neon, helium, krypton and xenon, because these are the rarest of all the elements on the Earth. Being very inert substances chemically, they do not combine with other elements to form compounds. With the exception of helium, which is continually being formed by the breaking up of the heavy radioactive elements, there is no reason to believe that these elements are present in the interior of the Earth; all that there is of them is contained in the atmosphere and how little there is of these elements in the atmosphere, argon alone excepted, may be seen from the table on p. 79, which gives the composition of the Earth's atmosphere. Some of these inert gases are known to have a high cosmic abundance. The amount of helium in the Sun is difficult to estimate with certainty but in the hotter stars, in whose spectra the lines due to helium are strong, helium is undoubtedly very abundant. It has been estimated that the abundance of helium on the Sun is about one hundredth of that of hydrogen ; the terrestrial abundance is probably not more than one ten-millionth as great. Neon lines are strong in the spectra of the nebulæ and the very hot stars, and it seems that the cosmical abundance of neon is some 500 million times greater than its terrestrial abundance. The discordance is not so marked for argon, whose cosmical abundance appears to be about equal to that of neon, whereas it is several hundred times more abundant on the Earth than neon. There is no information about the cosmical abundance of krypton and xenon.

Of the gaseous elements that are present in the atmosphere of the Earth the least discordance between the solar and terrestrial abundance is shown by oxygen. Though it is not easy to estimate accurately the amount of oxygen in the Sun's atmosphere, the available evidence suggests that oxygen may be about equally abundant on the Earth and on the Sun. Most of this oxygen occurs, however, not in the free state in the atmosphere but in chemical combination.

The similarity in the relative abundances of the metals in the Sun and Earth suggested a common origin. This common origin being accepted, the great differences in the relative abundances of the inert atmosphere-forming elements—nitrogen, argon, helium and neon—force us to the conclusion that these gases have been to a very great extent lost to the Earth. It is reasonable to suppose that when the Earth was first formed these elements were as abundant in the Earth as they are in the Sun and the stars. Being elements that do not readily enter into chemical combination with other elements, when the Earth began to cool and the temperature had fallen sufficiently for chemical compounds to form, these gases remained to a large extent in the Earth's atmosphere. But the Earth at that time was still very hot, so that it was losing its atmosphere at a very rapid rate, and the lighter gases were naturally being lost at a faster rate than the heavier gases. When it had cooled to such an extent that the loss of its atmosphere practically ceased, neon had been depleted to a much greater extent than the heavier argon. The atomic weight of neon is twenty, whereas that of argon is forty, as compared with atomic weights of fourteen and sixteen for nitrogen and oxygen respectively. In this way we have a plausible explanation of why argon is five hundred times more abundant than neon in the atmosphere of the Earth, whereas in the Sun and in the stars, neon is as abundant as argon.

The conclusion we have reached is that when the Earth had cooled sufficiently for the escape of its atmosphere practically to cease, almost all the neon that had been contained in the initial atmosphere had escaped. Much of the heavier argon had naturally escaped also, but the depletion of the argon was much less than that of the neon. Most of the original atmospheric oxygen, nitrogen and water-vapour, and almost all of the original helium and free hydrogen, must also have been lost, because these substances are lighter than neon. At this stage in its history, the Earth must have been almost devoid of an atmosphere.

As the molten Earth cooled still further, great quantities of water-vapour, carbon dioxide and other gases must have been evolved from the liquid magma, when at length it solidified to form the crust. These gases, with the residual gases from the initial atmosphere, formed the new atmosphere which, as the Earth was then relatively cool, could not escape. This atmosphere differed from the present atmosphere of the Earth in that it con-

PLATE 8

THE PLANET SATURN

In the upper portion of the plate is shown a photograph of Saturn obtained by Dr. Jeffers with the 36-inch refractor of the Lick Observatory on 1939, October 21. The outer ring has an exterior diameter of 171,000 miles and a width of about 10,000 miles; the inner ring has an outer diameter of about 145,000 miles and a width of about 16,000 miles. The width of the division between the rings is about 3,000 miles. The feebly luminous inner ring is not shown in the photograph. The shadow of the ball of Saturn on the ring will be noticed. The belts of Saturn are shown, and it will be seen that they are much less well-defined than the belts of Jupiter.

The lower portion of the plate shows photographs of Saturn taken in ultra-violet, violet, yellow, and red light. The differences in appearance are due to the different degrees of penetration of the light of different wavelength into the atmosphere of Saturn. The photographs were taken by Dr. W. H. Wright with the Crossley Reflector of the Lick Observatory, on 1929, August 25–26.

The different appearance of the rings at the two dates may be mentioned. In 1929 the rings were wide open, in 1939 they were viewed more obliquely. The appearance of the rings depends upon the elevation of the Earth above their plane, which can reach 27° as its maximum. When the Earth passes through their plane, the rings become almost invisible and appear like thin needles projecting from opposite si les of the planet.

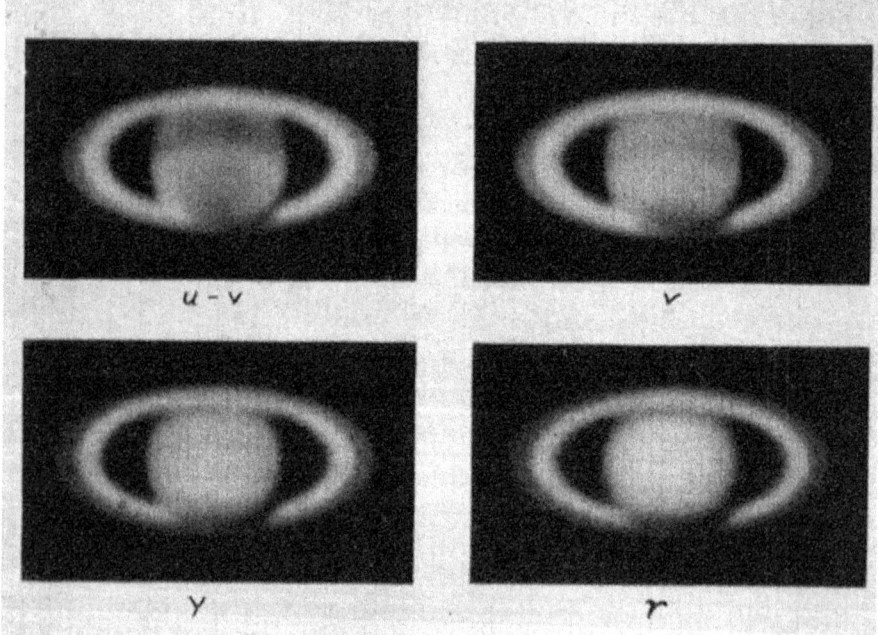

u - v   v

y   r

MOON, JUPITER, SATURN, URANUS and NEPTUNE

PLATE 9

THE SPECTRA OF THE MAJOR PLANETS

The plate shows in succession, from top to bottom, the spectra of the Moon, Jupiter, Saturn, Uranus and Neptune. The scale of wave-lengths is indicated at the top (the unit in which the wave-lengths are expressed is the " tenth-metre "—one-ten-thousand-millionth part of a metre—so that a wave-length of 5,000 is equivalent to ·00002 inch). The spectrum of the Moon is that of the sunlight reflected by the Moon. The absorption lines designated *A*, *B* by Fraunhofer are caused by oxygen in the atmosphere of the Earth; *a* is caused by moisture in the atmosphere of the Earth, *C* and *F* are caused by hydrogen in the Sun, *D* by sodium, *E* by iron and *b* by magnesium in the Sun. *A, a, B, C* are in the red, *D* is in the yellow, *E, b* are in the green and *F* is in the blue. These absorptions are present also in the spectra of Jupiter, Saturn, Uranus and Neptune. Other absorptions will be noticed in the spectra of these planets which do not appear in the spectrum of the Moon. These are produced by the atmospheres of the planets. In the spectrum of Jupiter, just to the left of *C*, is a weak absorption caused by ammonia. It is faintly visible in the spectrum of Saturn but is not seen in the spectra of Uranus and Neptune. Other absorptions increase in strength from the spectrum of Jupiter to that of Neptune; they are all produced by marsh-gas (or methane). They are so intense in the spectra of Uranus and Neptune that the yellow and red regions of these spectra are almost entirely cut out. It is for this reason that these planets show their characteristic green colour. A thickness of 25 miles of marsh-gas, at atmospheric pressure, is required to give absorptions as intense as those in the spectrum of Neptune.

tained a great amount of carbon dioxide and a great amount of water-vapour but not much oxygen. In course of time, as the Earth cooled still further, most of the water-vapour condensed out of the atmosphere and formed the oceans.

It remains to explain how the change from the atmosphere, as it then existed, to the present atmosphere has been brought about. For more than a century it has been recognised that the presence of free oxygen in the atmosphere, which we are apt to take for granted, demands explanation. Oxygen is an element that is chemically very active. Whereas the rare gases in the atmosphere are chemically inert and do not form compounds with other elements, oxygen does not like to exist alone. It is always eager to join up with other elements to form oxides. The rusting of iron is an illustration, rust being merely iron oxide. Combustion is nothing more than a process of oxidation, as we have already explained, and cannot occur in the absence of oxygen: if a glowing piece of wood is placed in a jar containing oxygen, it will at once burst into flame, because the oxidation then proceeds at a greatly enhanced rate.

Because of this preference of oxygen for combining with other elements rather than for existing alone, it must follow that processes are in continual operation that are depleting the store of oxygen in the atmosphere. One of the principal sources of depletion arises from the weathering of the igneous or basic rocks in the Earth's crust, a process leading to the formation of sedimentary deposits. The weathered material is carried down by streams and rivers and ultimately deposited on the sea floor as sand, clay or mud. The iron contained in the igneous rocks is not completely oxidised; the greyish hue of these rocks results from the iron being present mainly in the form of ferrous oxide, an oxide of iron in which the iron has not got the full complement of oxygen that it can hold. During the process of weathering fresh particles of rock are continually being exposed to the atmosphere and much of the ferrous oxide in the exposed parts becomes oxidised into ferric oxide, the red oxide of iron that is familiar to us as rust. It is the red ferric oxide that gives the characteristic red or brown colour to the weathered deposits.

The amount of oxygen that has been withdrawn from the atmosphere by this process is very considerable. The weathering process may be a slow one, but it must be remembered that the total thickness of the weathered deposits during geological times amounts to at least three or four hundred thousand feet. It has been estimated that the amount of oxygen that has been abstracted in this way from the Earth's atmosphere during geological times is equal to about twice the quantity now contained in the atmosphere. It is clear, therefore, that some other process must be in operation, which is continually replenishing the oxygen and making good the loss. The vegetation over the surface of the Earth provides the agency by means of which this replenishment is brought about. The green plant absorbs carbon dioxide from the air, as we have already seen in Chapter II, and uses the energy from sunlight to decompose it, the energy-transformer being the green colouring matter, chlorophyll, contained in the plant cells. The carbon is used to build up the complex organic substances found in living plants, the oxygen being returned to the atmosphere as a by-product.

We have, therefore, in continual operation, through the agency of the vegetation over the Earth's surface, the replenishment of the oxygen in the atmosphere at the expense of the carbon dioxide. The converse process is, however, also going on. Through the decay of vegetable matter and other organic materials, oxygen is absorbed and carbon dioxide is liberated. This carbon dioxide is again available for building up new plant cells. It may seem that we are arguing in a circle and that with two contrary processes in operation we have proved nothing and have got no nearer to explaining why the atmosphere now contains so much oxygen. This would be so if there were an exact balance between the two processes, but actually in geological times they have not balanced. Whenever organic matter is buried so that it cannot become oxidised and decay there is a net gain of oxygen to the atmosphere. Organic matter has in the past been buried on a large scale and has provided the coal measures and oil deposits of today. It seems probable that the present abundant supply of oxygen in the atmosphere has been provided at the expense of the carbon dioxide that it formerly contained, through the burial of the organic matter that now provides us with coal and oil. It has been estimated that if the coal, oil and other organic deposits could be unburied and completely burned, the whole of the oxygen in the atmosphere would be used up.

The Earth's atmosphere has thus passed through an evolution that is full of interest. The initial atmosphere, rich in hydrogen and helium, was largely lost whilst the Earth was still young and hot. A new atmosphere was provided as the Earth cooled. Still further cooling resulted in the formation of the oceans. Then, through the action of vegetation, there has been the replacement of carbon dioxide by oxygen; the change from an atmosphere rich in carbon dioxide to an atmosphere rich in oxygen is accounted for by the burial of large quantities of vegetation, which have thereby been preserved from decay.

This study of the evolution of the atmosphere of the Earth will be helpful when we come to investigate the atmospheres of the other planets. It suggests that the values of the velocities of escape at the present time will provide a guide to an estimate of the *maximum* amount of atmosphere that a planet is likely to have; we have obtained definite evidence that the Earth must have lost much of its initial atmosphere at a rapid rate whilst it was still hot; the possibility of this initial rapid loss must be borne in mind in the case of the other planets. We have learnt also that an abundance of oxygen and a scarcity of carbon dioxide is indicative of plentiful vegetation, whereas, in the absence of vegetation, we are likely to find an abundance of carbon dioxide and a scarcity of oxygen.

CHAPTER V

WORLDS WITHOUT ATMOSPHERES

WHEN the velocity of escape from a planet is sufficiently small, the atmosphere will have been dissipated at a rapid rate, so that we should not now expect to find any traces of an atmosphere. These conditions are likely to be found on the small bodies of the solar system; for the velocity of escape is determined by the ratio of the mass to the radius of the planet, and though the mass is small for the planets of small size and large for the planets of large size, the mass is dependent upon the cube of the radius; it follows that, apart from differences in mean density from one planet to another, the velocity of escape is proportional to the radius of the planet. It is, therefore, the small planets that will have the low velocities of escape. The figures given on p. 61 confirm this.

Of the bodies listed there, the Moon has the lowest velocity of escape, viz. 1·5 miles a second. The observed maximum temperature on the Moon is 120° C.; at this temperature the average molecular speeds are twenty per cent. greater than the speeds at the temperature of 0° C., which are given for a selection of gases on p. 56. The criterion for the retention of an atmosphere tells us that at its present temperature the Moon could retain carbon dioxide and any heavier gases, but oxygen and all lighter gases—including nitrogen, water-vapour, helium and hydrogen—would be lost. In the earlier stages of the Moon's history, when it was much hotter than it now is, the rate of escape of atmosphere must have been very rapid. Even at the present time the Moon would lose an atmosphere of hydrogen almost instantly. At a temperature of 1,000° C. it would lose an atmosphere of carbon dioxide in a few years. We may expect, therefore, to find that the Moon is now totally devoid of any atmosphere.

The next lowest velocity of escape is that for Mercury, 2·4 miles a second. The observed average temperature of the sunlit side of Mercury is about 400° C., and at this temperature the average molecular velocities have 1·57 times their values at 0° C. At such a temperature, Mercury could retain atmospheres of carbon dioxide and oxygen but not of lighter gases. But since the maximum temperature on the sunlit side of the planet is appreciably higher than 400° C., the chance of retaining an atmosphere is less favourable than this suggests. If Mercury had remained for long at much higher temperatures before it cooled to its present state, it must also, like the Moon, have lost its atmosphere entirely. We can have little hope of finding any evidence of an atmosphere on Mercury.

We come next to Mars, from which the velocity of escape is 3·2 miles a second and whose temperature is rather lower than that of the Earth. Mars could not at the present time retain either hydrogen or helium in its atmosphere, but it can retain water-vapour and heavier gases. Admitting a rapid loss of atmosphere in its early years, when its tempera-

ture was high, we must nevertheless conclude that
Mars is likely to have retained a certain amount of
atmosphere. The conditions are, however, somewhat critical, and Mars is barely able to retain an
atmosphere. It is a planet of particular interest
and will be considered in detail in a later chapter.

We have not hitherto made any mention of the
satellites of planets other than the Earth. Mars has
two satellites, Jupiter has eleven, Saturn has nine,
Uranus has four and Neptune has one. Are any of
these at all likely to possess an atmosphere? They
differ considerably in size; some, including the two
satellites of Mars and the smallest satellites of
Jupiter's large family, are not more than a dozen or
so miles in diameter, whilst the largest slightly
exceed the planet Mercury in size. It is certain
that, with the possible exception of the few largest
satellites, it is quite impossible for them to have
retained any atmosphere at all.

Particulars of the largest satellites are as follows:

Planet.	Satellite.	Diameter (Moon = 1).	Mass (Moon = 1).	Velocity of escape miles/sec.
Jupiter . .	Io	1·07	1·09	1·5
Jupiter . .	Europa	0·91	0·65	1·3
Jupiter . .	Ganymede	1·48	2·10	1·8
Jupiter . .	Callisto	1·49	0·58	0·9
Saturn . .	Titan	1·21	1·86	2·0
Neptune . .	(No name)	1·4 ?	?	?

The satellite of Neptune is too far away for its size
to be known with any accuracy, but from its brightness it is estimated that it is not much inferior in

size to Ganymede and Callisto, the largest of the satellites of Jupiter. Nothing is yet known about its mass.

The velocities of escape from these satellites, given in the last column, may be compared with the velocity of escape from the Moon, 1·5 miles a second, and from Mercury, 2·4 miles a second. For three of the satellites the velocities of escape are equal to or are less than the velocity of escape from the Moon; for the other two, the velocities are intermediate between the velocities of escape from the Moon and Mercury. The temperatures of these satellites are all very low, so that it would be possible for them now to retain atmospheres of the heavier gases—though gases lighter than water-vapour could not be retained. But they will have passed through a high-temperature stage, and if this stage had lasted for some time they must have lost their atmospheres entirely. If any of them has an atmosphere at the present time it must be extremely tenuous.

The conclusions that we have reached about the atmospheres of the Moon and of Mercury have been based entirely on theoretical considerations. It is of interest to inquire whether these conclusions are confirmed by observation. The Moon is our nearest neighbour in space and can be studied in greater detail than any other celestial body. Its distance is only about a quarter of a million miles. Such a distance may seem large when we compare it with distances on the Earth; an aeroplane travelling without any stop at a speed of two hundred miles an hour would take seven weeks to reach the Moon. But, in comparison with most of the distances that are the concern of the astronomer, the Moon can properly be described as a near neighbour. It is so near that an object on the Moon of the size of St. Paul's Cathedral could just about be detected under favourable conditions.

In the telescope the Moon appears as a rugged mountainous world. There are great mountain ranges on the Moon, the finest being the Apennines, shown in Plate 4, but the most prominent and characteristic feature of its surface is the great number of mountain rings, which cause large areas of the surface to have a sort of pock-marked appearance. We are able to see only one half of the surface of the Moon, because the Moon always turns the same face towards the Earth, the other face being permanently turned away from us. This condition has been brought about through the action of the gravitational attraction of the Earth. When the Moon was young and its surface was still plastic, the attraction of the Earth raised a great tidal bulge on its surface; as the Moon rotated, this tidal bulge was dragged across it so as always to face towards the Earth. The dragging action formed a brake on the rotation of the Moon and gradually slowed it down until at length the Moon turned always the same face towards the Earth. An analogous effect is still slowing down the rotation of the Earth. The tides in the oceans, produced by the gravitational action of the Moon, act as a brake upon the Earth, causing a very slow but progressive lengthening in the day. The magnitude of the effect is small, the length of the day increasing in the course of a century by only two-thousandths of a second; but the effects are cumulative and will continue until eventually the Earth will always turn the same face to the Moon. When that happens the length of the day will have increased to about forty-seven of our present days.

The ring mountain formations on the Moon are commonly called *craters*, from their resemblance to the volcanic craters on the surface of the Earth, though they are on a much greater scale than the largest terrestrial craters. The largest lunar craters have a diameter of more than one hundred miles, but there are many others that do not measure more than a mile or two across. Small craters are often found within larger ones and, in many cases, the wall of one crater breaks through the wall of another, suggesting that they were formed at different times. The complexity of the topography of the Moon, and the extreme ruggedness of its surface, are well illustrated by the portion of the surface shown in Plate 5. The designation of these ring formations by the term *crater* is perhaps misleading, because it suggests a definite volcanic origin, whereas it is by no means certain that they have been formed by volcanic action. One theory of their origin does suppose, however, that they are the result of volcanic activity. Many of the craters have a central peak and it is suggested that material was ejected from a central volcano and gradually built up the ring-shaped wall. The force of gravity is much smaller on the Moon than on the Earth and this would permit matter to be ejected without difficulty to great distances. There is, however, no essential difference in general structure between the largest and the smallest craters, and it is difficult to understand how the largest craters, one hundred or more miles in diameter, could have been produced by volcanic action. Evidence of past volcanic activity on the Moon seems to be provided, nevertheless, in another way, by the great plains or " seas " as they are commonly called; it is probable that these were once actual seas of lava and irregular ripple-like markings on them appear to indicate the limits of successive waves of lava flow.

A different theory of the craters supposes that they were formed as the result of the bombardment of the Moon by gigantic meteors. There are a few examples of meteor bombardment to be found on the Earth's surface, the largest being the Great Meteor Crater in Arizona, a hollow about 600 feet deep and rather more than a mile in diameter, with a raised rim. It is possible that in the early stages of the Earth's history there was much more material, which had not been aggregated into planets, in the neighbourhood of the Sun than there now is, and that both the Earth and the Moon were subjected to intense meteoric bombardment. The Moon still bears the marks of this bombardment in its much-scarred surface; but the action of denudation by water and erosion by wind, with the formation of sedimentary deposits, and successive upheavals to form mountain ranges have long ago wiped away all the traces from the Earth's surface of the bombardment that it had once undergone. The few meteor craters that can be seen on the

Earth at the present time are formations that are geologically young.

Other formations of interest may be seen on the surface of the Moon. There are many straight clefts or crevasses, about half a mile or so in width, which run in more or less straight lines for hundreds of miles through valleys and across mountains. They appear to be cracks in the surface, caused by the contraction of the interior of the Moon. There are also many deep narrow valleys and some straight lines of cliff caused by faulting of the surface. The most puzzling of all the features are the light-coloured streaks or rays, which radiate radially from some of the craters and pass across valleys and mountains and over craters for distances of hundreds of miles. They are a few miles in breadth and do not appear to be elevated above the surface. The true nature of these rays is not known, through it has been suggested that the appearance is caused by the staining of the surface by vapours coming from narrow rifts. The rays from the great crater Copernicus are shown in Plate 6.

The appearance of the Moon in the telescope, with its steep rugged mountains, jagged rocks and fault scarps, suggests a world where there has been no wearing away by erosion. It is quite certain that the Moon is a waterless world. Oceans, lakes and rivers would be clearly seen if they existed and at times they would reflect the sunlight and appear intensely bright. No clouds ever veil the Moon's surface. This is merely what we should expect if, as we have concluded, the Moon has no atmosphere.

If there were any water on the Moon it would rapidly evaporate during the heat of the long lunar day and the water-vapour would be dissipated away into space.

The absence of an atmosphere is indicated also in other ways. The telescopic appearance of the Moon itself proves that there cannot be more than the merest trace of atmosphere, for the parts of the Moon near the edge of the disk are as sharply defined as the other parts of the surface; if there were any atmosphere, the edge portions of the Moon would be seen through a much greater depth of atmosphere than the central portions and would be partially or even completely obscured. Further evidence is provided by observing the occultation of a star as the Moon passes in front of it. The Moon moves eastward across the sky relative to the stars; between New Moon and Full Moon the advancing edge is the dark one. When the dark edge passes in front of a star, the star retains its full brightness until it instantaneously disappears. There is no gradual fading away as there would be if the Moon had any atmosphere; at one moment the star is seen to be shining with its usual brightness; the next moment the star is not there. It is as though it had suddenly been completely blotted out of existence. And however often the observation is made, the suddenness of the disappearance never ceases to cause surprise. Between Full Moon and New Moon the advancing edge is the bright one; a star that has been occulted then reappears with equally startling suddenness as the dark edge passes away from it.

We see the Moon by means of sunlight falling upon it which is reflected back by its surface. At any time one half of the surface is in sunlight and the other half in darkness. Near New Moon the sunlight is mostly falling on the hemisphere of the Moon that is turned away from us; near Full Moon it is mostly falling on the hemisphere that is turned towards us. This is the cause of the varying phases of the Moon. Though the Moon appears very bright, its surface is actually a poor reflector; less than ten per cent. of the sunlight that falls on it is reflected back, the remainder being absorbed and going to heat the surface. The reflecting power of the surface of the Moon corresponds to that of greyish-brown rocks. It may be remarked that the greyish colour of the Moon is suggestive of rocks that are incompletely oxidised and indicate the absence of oxygen; if oxygen were present on the Moon to oxidise the rocks completely, the colour of the Moon would be reddish, somewhat resembling the colour of Mars.

The temperature of the Moon can be measured with the aid of a bolometer or thermocouple, as previously described. As we have just mentioned, most of the radiation that we receive from the Moon is merely reflected sunlight. The remainder is heat radiation—radiation that has been absorbed by the surface rocks, which become heated and radiate like a brick wall that has been heated in the Sun. This heat radiation, characterised by its long wavelength, can be separated from the reflected sunlight, which is mainly of much shorter wave-length, and used to determine the surface temperature. During the long lunar day, equal to fourteen of our days, the surface rocks become very hot and the temperature near the equator of the Moon at noon is about 120° C., considerably higher than the temperature of boiling water at the surface of the Earth (100° C.). The temperature falls rapidly as the altitude of the Sun decreases and by sunset it is already freezing hard, the temperature being then −10° C. (or 18° of frost on the Fahrenheit scale). The most rapid variations of temperature occur at the time of an eclipse of the Moon. As the Earth advances, throwing its shadow on the Moon and cutting off the sunlight from its surface, the temperature immediately begins to drop with startling rapidity. At one eclipse, for instance, it was found that the temperature of the surface dropped from +70° C. to −80° C. in a little more than an hour. During totality, which lasted 2½ hours, the temperature dropped a further 40° C. But after totality had ended, the temperature rose in an hour to almost its initial value. These extreme and rapid variations are what we should expect on a world that is entirely devoid of atmosphere. We may compare these figures with the fall of temperature experienced on the Earth at points within the belt of totality on the occasion of a total eclipse of the Sun; the fall of temperature does not usually exceed two or three degrees.

From what we have learnt of the Moon, and in particular from its lack of oxygen and water and from its extreme variations of temperature, we should naturally conclude that it is a world where life of any sort is entirely out of question. The

Moon is the only world where we should expect actually to see clear evidence of life, if any existed. There is no doubt that if there were a lunar inhabitant equipped with a powerful telescope, he would be able to see many signs of human activity on the Earth. He would be able to watch the growth of greater London; he would see cities like New York, Sydney, Johannesburg and Ottawa springing up. He would be able to watch the formation of new lakes by the impounding of water by dams. He would see land-reclamation works in progress and the draining of the water from tracts such as the Zuyder Zee. The seasonal growth of vegetation and the melting of the snow over vast tracts of land with the advance of summer would be clearly visible to him. In the course of a few years he would undoubtedly obtain clear evidence not only of plant life on the Earth but also of human activity.

So, in a similar way, if the Moon were inhabited by intelligent beings we should expect to find plenty of evidence of their existence. We can find none. The Moon shows no signs of change. There are not even seasonal changes of colouration, such as might be attributed to the growth of vegetation. Some astronomers have claimed, indeed, to have seen slight changes in certain regions, changes mainly of tint such as might be produced by the growth of lichens on rocks. But such changes have not been confirmed and are generally discredited. It seems that the observers have been misled by changes in the appearance of the surface detail with changing altitude of the Sun. No! it is not possible to admit that there is life of any sort on the

8

Moon.   It is a world that is completely and utterly dead, a sterile mountainous waste on which during the heat of the day the sun blazes down with relentless fury, but where during the long night the cold is so intense that it far surpasses anything ever experienced on the Earth.

These hard facts are conveniently ignored by those who believe that it would be possible to shoot a rocket, containing human beings, to the Moon from which the human explorers could land and explore some portion of the Moon's surface.   The explorers would need to be encased in airtight suits and provided with oxygen apparatus to enable them to breathe.   Even supposing that they could protect themselves against the great heat by day and the extreme cold at night, a worse fate might be in store for them unless their suits were completely bullet proof.   For they would be in danger of being shot by a shooting star.   The average shooting star or meteor, which gives so strongly the impression of a star falling from the sky, is a small fragment of matter, usually smaller than a pea and often no larger than a grain of sand.   Space is not empty but contains great numbers of such fragments.   The Earth, in its motion round the Sun, meets many of these fragments, which enter the atmosphere at a speed many times greater than that of a rifle bullet. The meteor, rushing through the air, becomes intensely heated by friction and is usually completely vaporised before it has penetrated within a distance of twenty miles from the surface of the Earth.   Many millions of these fragments enter our atmosphere in the course of a day, but the atmosphere protects us from them. On the Moon, however, they fall to the surface and so great is their number that the lunar explorers would run a considerable risk of being hit.

The difficulties that would have to be encountered by anyone who attempted to explore the Moon—assuming that it was possible to get there—would be incomparably greater than those that have to be faced in the endeavour to reach the summit of Mount Everest. In two respects only would the lunar explorer have the advantage. In the first place movement would be less fatiguing because—as the weight of the Moon is only about one-eightieth of that of the Earth—the gravitational pull of the Moon is not very great. If the Moon had an atmosphere like that of the Earth, a golfer on the Moon would find that he could drive his ball for a mile without much difficulty and a moderate batsman would hit sixes with the greatest of ease and perform feats that even Bradman might envy. The second advantage the lunar explorer would have over the climbers on Mount Everest would be the absence of strong winds to contend against. The Moon having no atmosphere, there can be no wind; nor, of course, can there be any noise, for sound is carried by the air. The Moon is a world that is completely still and where silence, " a silence where no sound may be," prevails.

We turn next to the planet Mercury, the nearest of the planets to the Sun. The year on Mercury, the time that is taken for Mercury to traverse its orbit around the Sun, is 88 of our days. The orbit of Mercury is much the most elliptical of any of the planets,[1] so that the distance of Mercury from the Sun varies rather widely in the course of its passage around the Sun.   It ranges from about 28½ million miles, when at its nearest, to about 43½ million miles when at its greatest distance.   Mercury is the fastest moving of all the planets, its speed ranging from thirty-six miles a second, when it is at its nearest to the Sun, to twenty-four miles a second, when farthest away.   It is, therefore, not inappropriately named after the winged-footed messenger of the ancient gods.

Mercury turns always the same face towards the Sun, just as the Moon turns always the same face towards the Earth.   In the course of its journey round the Sun, it therefore makes exactly one rotation about its axis.   In other words, the length of the day (defined as the time taken by the planet to make one rotation about its axis) is equal to the length of the year.   But the other use commonly made of the word *day*, as contrasted with *night*, ceases to have any meaning on Mercury.   For a portion of the planet has perpetual day and another portion has perpetual night.   If Mercury were always at the same distance from the Sun, exactly half of the planet would have perpetual day-time and the other half would have perpetual night-time.   This condition is modified somewhat by the variation in the distance of Mercury from the Sun.   About three-eighths of the surface has perpetual day and an equal area has perpetual night.   In the

[1] We leave out of consideration the thousands of minor planets or asteroids which circulate round the Sun between the orbits of Mars and Jupiter.

intermediate zone, embracing about one-quarter of the surface, the Sun alternately rises above the horizon and falls below it.

The temperature of the portion of Mercury that is perpetually bathed in sunlight is very high, about 400° C., a temperature considerably higher than that at which lead melts and close to the temperature required for zinc to melt. The opposite side, which never receives any sunlight at all, must be intensely cold. In the zone between, which has alternately sunlight and darkness, the variations of temperature must be extremely great.

We have seen that there is little expectation of any atmosphere on Mercury. Though, at the present time, Mercury could hold an atmosphere composed of the heavier gases, the atmosphere must have escaped entirely if Mercury had remained very hot for any length of time after its formation. We have already concluded that the Earth, where conditions are much less favourable for the escape of an atmosphere, must have lost most of its original atmosphere whilst it was still hot. It is just possible that though Mercury lost all its original atmosphere, gases were given off by the crust as it solidified and, if so, Mercury may possibly now have an extremely tenuous atmosphere of carbon dioxide.

It is not easy to find out anything very definite about conditions on Mercury by direct observation. Being near the Sun, Mercury is never visible for very long after sunset or before sunrise and it is comparatively seldom seen with the naked eye. The great astronomer, Copernicus, died without ever having seen it; he did not have the advantage of the optical aid provided by telescopes, which makes it possible for Mercury to be seen in broad daylight. But broad daylight does not usually provide favourable conditions for attempting to see any markings on the surface of Mercury.

Exceptionally steady atmospheric conditions are required for the study of the surface of Mercury. In 1882 an Italian astronomer, Schiaparelli, began to make a systematic study of this planet, and in 1889 he announced that there were markings on the surface, which were of a permanent nature and which indicated that Mercury always turned the same face to the Sun. The markings were ill-defined and gave the surface a spotty appearance. Subsequent observations by other observers have confirmed Schiaparelli's observations in their broad essentials, though there are considerable differences between the markings recorded by different observers. This is not surprising, for the observations are of extreme difficulty. There is, nevertheless, general agreement about the main features, and it seems reasonably certain that the markings are due to irregularities of the surface—as in the case of the Moon. It is possible that the surface of Mercury bears a general resemblance to the mountainous surface of the Moon.

Difficult as it is to observe the surface markings of Mercury, it is far more difficult to obtain any certain evidence of an atmosphere. The reflecting power of Mercury, is about equal to that of the Moon. If there was much atmosphere, Mercury would undoubtedly reflect a good deal more of the sunlight that falls upon it. Observations of the changes of

brightness of the planet with changes of phase
suggest that there can be little or no atmosphere,
and that the light is reflected from a rough surface,
like that of the Moon.

It sometimes happens that Mercury passes
directly between the Earth and the Sun; it may
then be seen projected on the Sun's disk as a small
round black spot. The moments when it is entering
on to, or leaving, the disk of the Sun provide
favourable opportunities for detecting any atmosphere. At such times, if there were an appreciable
atmosphere on Mercury, the portion of its disk
outside the Sun would be surrounded by a bright
line or halo, produced by the refraction of sunlight
through the planet's atmosphere. Such a bright
halo is seen when Venus is entering upon the Sun's
disk, but it is not noticed in the case of Mercury.
This observation provides definite evidence that,
if there is any atmosphere on Mercury, it must be
of extreme rarity.

Some observers have felt confident, nevertheless, that Mercury must have a thin atmosphere
because, they assert, certain of the surface markings
are obscured from time to time. The temperature
is too high for any sort of condensation cloud to form
above the surface and it has accordingly been
suggested that the obscuration is caused by dust.
It is possible that there may be active volcanoes on
Mercury which eject clouds of dust to a great
height at times of violent eruptions, and that there is
a very thin atmosphere, sufficient to prevent the
fine dust settling too rapidly. More will be
learned in the future, we may hope, about this.

The observations are extremely difficult and are not free from doubt. They need further confirmation. At present all that we can positively assert is that Mercury can have at the most an atmosphere of extreme rarity, and that the existence of such an atmosphere is not impossible on theoretical grounds.

Whatever the final conclusion about the atmosphere of Mercury may be, there seems to be no possibility that life can exist on the planet. The high temperature over much of its surface, the very low temperature over other parts and the extreme variations of temperature over the intermediate zone, together with the absence of water-vapour and oxygen—which must have been lost whilst Mercury was much hotter than it is at present—combine to make conditions under which no form of life could survive.

The two largest satellites of Jupiter, Ganymede and Callisto, are larger than the planet Mercury; the other two major satellites of Jupiter, Io and Europa, the largest of the satellites of Saturn, Titan, and the satellite of Neptune though smaller than Mercury are larger than the Moon. The velocities of escape from these bodies do not differ greatly from the velocity of escape from the Moon, and at their present temperatures they could just about retain water-vapour and heavier gases and vapours. But it is not possible to suppose that they have been very cold throughout their history and it is probable that their atmospheres were entirely or almost entirely lost whilst they were young and hot.

Surface markings can be seen on some of these bodies and they all appear to turn the same face continually to the parent planet. They all have rather low mean densities; Io and Europa have mean densities nearly three times that of water; these two satellites are probably masses of rock like the moon. The mean density of Gannymede is slightly more than twice that of water, whilst that of Callisto is only six-tenths that of water. These two bodies may consist largely of rock with a thick covering of ice or solid carbon dioxide. The very low mean density of Callisto may be spurious, as the mass of this satellite is difficult to determine and is somewhat uncertain; if real, it is difficult to explain unless we may suppose that its rocks are mainly in the form of pumice.

The average temperatures of the satellites will not differ much from those of the parent planets because a planet and its satellites are at approximately the same distance from the Sun. Reference to the table of temperatures on p. 76 shows that the large planets are extremely cold. These low temperatures would render the development of any form of life impossible. We may conclude that it is unlikely that any of the satellites in the solar system have an atmosphere; but in those cases where there may remain some doubt about the correctness of this assertion, the temperatures are certainly too low for the existence of life. Hence we can leave all the satellites out of consideration in our search for conditions in the solar system where life might conceivably exist.

CHAPTER VI

THE GIANT PLANETS

THE four major planets, Jupiter, Saturn, Uranus and Neptune, are much larger and much more massive than the Earth. Their radii and masses, relative to the corresponding quantities for the Earth, are given in the table on p. 61. The same table gives the velocities of escape from these bodies. These velocities are all high and are, in fact, so much greater than the average molecular velocity of hydrogen that it is impossible for hydrogen, and therefore also for any of the heavier gases, to have escaped from their atmospheres, even if we suppose that they were initially much hotter than they now are. The giant planets must consequently have retained their initial atmospheres in their entirety. We may accordingly expect to find that they all possess dense and extensive atmospheres. For this reason they have many similarities, and it is convenient to consider them together.

The nearest and the largest of these four planets is Jupiter, which is, therefore, the most favourably placed for detailed observation and study with the telescope. A small telescope is sufficient to show Jupiter as a fine-looking object. The *apparent* angular diameter of Jupiter varies from thirty-two seconds of arc, when it is at its greatest distance from the Earth (600 million miles), to fifty-two seconds when it is at its nearest (367 million miles). Even when it is at its greatest distance, a low magnification of sixty will make it appear in the telescope as large as the Moon appears to the naked eye. We suppose, then, that we are looking at Jupiter in a moderate-sized telescope. We see a bright disk, crossed by a number of dark markings arranged in more or less parallel belts. The disk is not circular in outline, but is distinctly flattened, the shortest axis being at right angles to the parallel belts. This flattening suggests a body that is in rapid rotation. If the flattening is produced by rotation, the shortest axis will be the axis of rotation; the belts are therefore parallel to the equator of Jupiter. When we look attentively at the dark markings, we notice that they are not of a uniform structure and of sharply defined outline but that they show great complexity of structure and variety of detail. Fixing attention on any convenient well-defined object on the surface of the planet, we notice that it appears to move gradually across the surface. This apparent motion is produced by the rotation of Jupiter on its axis. If we watched for a sufficiently long time, we should see the object disappear at one limb of Jupiter and after several hours reappear at the opposite limb; in rather less than ten hours from the commencement of the watch we should find that the object had returned to the same position on the disk that it occupied when the observations commenced. We thus confirm the rapid rotation of Jupiter, which the flattened shape of the planet had caused us to suspect, and we find that the time of rotation is rather less than ten hours. Jupiter has, in fact, the shortest period of rotation of any of the planets.

If our observations are made in the twilight, before the sky has become dark, we will notice that the centre of the disk is much brighter than the edge. This effect is much less noticeable at night, when the sky is dark, because the contrast in brightness between the limb and the dark sky then makes the edge portions of the planet appear subjectively relatively brighter than they really are. Precise measures of the brightness with a photometer show that the brightness begins to fall off rapidly near the edge of the disk and that at the edge the brightness is only about one-eighth that at the centre. This suggests that Jupiter has an atmosphere and that the falling off in brightness towards the edge is produced by the absorption of light in this atmosphere. In the case of the Sun, which is a gaseous body, there is a similar decrease in brightness from the centre to the edge of the disk, whereas the Moon, which is a solid body without an atmosphere, does not show any difference in brightness between centre and edge.

If we were to embark on regular systematic observations of Jupiter we should soon find that the phenomena shown by the surface markings are of considerable complexity. It was noticed as long ago as the seventeenth century that the rotation period was not the same for all portions of the planet, the equatorial regions rotating faster than the polar regions. In other words, Jupiter is not spinning round on its axis like a solid body; this provides further evidence that the visible layers are not solid but that they must be gaseous or, possibly, liquid. A similar lack of uniformity in its rotation is shown by the Sun, whose equatorial regions rotate considerably more rapidly than the polar regions. The period of rotation of the Sun at the equator is about 24⅔ days, whilst at the poles it is about 34 days. But whereas the period of rotation of the Sun increases progressively from equator to poles, there is no corresponding regularity in the case of Jupiter. There are a number of different zones, each of which is pretty sharply defined and each of which has its own rate of rotation, with abrupt discontinuities from one zone to the adjacent zones.

The pioneer work in the systematic detailed study of these movements was made by a British amateur astronomer, Mr. Stanley Williams, who died in 1938. He established the existence of a number of definite currents in the surface material; these currents determine the rotation periods of the spots or markings that appear within them. Williams showed further that the currents were of a permanent nature, although from time to time there were temporary variations in the rates of their movement. The results obtained by Williams gave a great impetus to the study of the surface markings of Jupiter and the planet has been kept under systematic observation by the devoted band of amateur observers that forms the Jupiter Section of the British Astronomical Association, pre-eminent amongst whom is the Rev. T. E. R. Phillips, who directed the work for more than thirty years. Two drawings of Jupiter by Mr. Phillips are reproduced in Plate 7, along with two photographs taken with the 36-inch refracting telescope of the Lick Observatory.

These observations have confirmed the perman-
ence of the eleven currents found by Williams.
The most important of these is the great equatorial
current, which covers a zone from 10,000 to 15,000
miles wide and has a period of rotation of rather
more than 9 hours 50 minutes. The periods of
rotation of the other currents lie between 9 hours
55 minutes and 9 hours 56 minutes, but bear no
relation to the latitude. The distribution of the
currents is different in the two hemispheres. The
differences in period of rotation of the several currents may seem to be not very large and they are,
of course, very much smaller than the differences in
the rates of rotation of the various portions of the
Sun. But the differences are perfectly distinct and
definite, so that well-marked features in adjacent
currents may drift past one another, as the result of
the different rates of movement of these currents, at
a relative speed of as much as two hundred miles
an hour.

Though the different currents are permanent,
their positions as well as their rates of rotation may
vary from year to year. These variations have been
well established by the observations of Williams,
Phillips and others and are not to be attributed to
errors of observation. The movement of the currents and of the spots or other markings associated
with them is usually almost entirely in a direction
parallel to the equator, though from time to time
spots have been observed whose movement is of a
circulatory nature.

Most of the markings to be seen on Jupiter are
short lived. Some are merely transient, others may

laṡt for a few weeks or, occasionally, for a few months. Their shapes show continual change. There can be little doubt that they are atmospheric phenomena of some sort. Their appearance suggests that they may be clouds, formed of droplets of condensed vapours, floating in the atmosphere of Jupiter.

There are a few markings, however, that are of a much more permanent nature. The most remarkable of these is the great Red Spot. This marking was seen in 1878, and has been visible ever since, though its appearance changes so much that at some times it is very conspicuous and of a striking brick-red colour, whilst at other times it loses its colour and appears to fade away until it can scarcely be seen. It is normally oval in shape, extending some 30,000 miles in length and with a width of about 7,000 miles, the long axis being parallel to the equator; from time to time the shape changes and it becomes rounder. From a study of earlier drawings and observations of Jupiter it has been concluded that it was observed as far back as 1831 and that it may well be identical with a marking observed by Hooke in 1664. The appearance in 1878 came after a period of relative faintness. It was at one time thought that the Red Spot was a portion of the surface of Jupiter seen through a break in the clouds or that it was in some way attached or related to the solid surface; it has been suggested, for instance, that it might be a cloud or pall of smoke and dust hanging over an active volcano on Jupiter. But it has been found that the period of its rotation is subject to large and irregular variations, so that it is impossible that it can be related to the solid surface of the planet.

Another remarkable marking is known as the South Tropical Disturbance. It is a dark region, some 45,000 miles in length, lying in the belt just south of the Red Spot, which was first seen in 1901. Its rotation is somewhat more rapid than that of the Red Spot, so that it overtakes the Spot at intervals which were at first about two years in length but which have gradually become longer, streaming past it at a relative rate of several miles an hour. When the disturbance is overtaking the Spot, its motion becomes accelerated and, in passing the Spot, it tends to drag the Spot along with it for several thousands of miles, the Spot then drifting back to its previous position.

The belts on Jupiter are rich in colour, the colours being sometimes very vivid. Though red, brown and orange colours predominate, olive green and bluish patches may also be seen. The colouration of the belts is variable but the two hemispheres of the planet usually vary in an opposite manner. When the belts in one hemisphere show a maximum of redness, those in the other hemisphere are colourless or slightly bluish and conversely, whilst at intermediate times the predominant effect in both hemispheres is a moderate redness.

It was thought until within fairly recent years that Jupiter was almost red-hot and that the bright colours seen in the belts were due to glowing vapours rising into the atmosphere from the almost molten mass. It was known, however, that Jupiter was not sufficiently hot to be perceptibly self-luminous; for from time to time one or other of the satellites passes between the planet and the Sun and, when this happens, the shadow of the planet is cast upon the surface; these shadows appear quite black, which they would not do if there was any appreciable glow from the surface. Observations within the last twenty-five years have proved that the belief that Jupiter is hot was ill-founded. Direct measurements have shown that most of the radiation that reaches us from Jupiter is merely reflected sunlight; the small portion that is true planetary radiation indicates that the temperature of the surface is very low, about $-140°$ C. This temperature is in close agreement with the temperature that would be expected for a planet at the distance of Jupiter from the Sun, if the surface were warmed only by the radiation from the Sun. It can be inferred that there is a close balance between the radiation that Jupiter receives from the Sun and the radiation that it re-emits to space, leaving little or no residuum to be accounted for by heat coming from the interior of the planet.

The appearance of Saturn in the telescope, leaving the rings out of consideration, is generally similar to that of Jupiter. The surface is marked by bright and dark belts parallel to the equator. But whereas the belts of Jupiter are very irregular in outline, show a great deal of detail in the form of bright and dark spots, and undergo rapid changes, the belts of Saturn are regular in outline, are ill-defined and show little detail and have few spots or other irregularities. Information about

9

the rotation period of Saturn and about the variation of the rotation with latitude is therefore not readily obtainable. From time to time, however, white spots appear, which may persist for some time. In 1794 Sir William Herschel observed such a spot and was able to determine its period of rotation, which he found to be ten hours sixteen minutes; the rotation of Saturn is therefore slightly slower than that of Jupiter. Other spots have been observed from time to time, at rather infrequent intervals. The most conspicuous spot ever observed on Saturn appeared in 1933 and was first seen by Mr. Will Hay in London. This spot, like the spot observed by Herschel in 1794, was near the equator and the rotation period of the two spots were in close agreement. As with Jupiter, the higher the latitude the slower is the rotation, though the change with latitude is much greater for Saturn than for Jupiter. A spot discovered by Barnard in 1903 in latitude 36° north gave a period of ten hours thirty-eight minutes.

As in the case of Jupiter, the edges of the disk of Saturn are much less bright than the centre, suggesting the presence of a considerable atmosphere. The disk is also strikingly flattened, the axis of rotation being the short axis. The flattening is more marked than for Jupiter; Saturn is, in fact, much the most flattened of all the planets. This is another indication of an extensive atmosphere. Saturn shows colours which are not so strongly marked as those of Jupiter. The equatorial zone is usually of a bright yellow colour and there is a darkish cap, of a greenish hue, at the pole. The measured temperature of Saturn is very low, $-155°$ C., some $15°$ C. lower than that of Jupiter, as would be expected from its greater distance from the Sun.

It is the rings of Saturn that make it such a unique and striking object in the telescope. In the discussion of conditions on Saturn, in relation to the possibility of the existence of life on the planet, we are not really concerned with the rings, which are not solid and could not be the home of life. It was shown in 1857 by Clerk Maxwell that the rings could not continue to exist unless they were composed of a multitude of small particles: if the rings were solid, liquid or gaseous they would be unstable and would break up. The rings may therefore be considered as consisting of a great number of tiny moons, circulating around Saturn. The rule of the road that governs such motion under the controlling force of gravitation is that the outer lanes of traffic move more slowly than the inner lanes. That this is actually the case was proved by observation in 1895 by Keeler. The rings naturally do not show any marks that can be identified and used to serve as a measure of the rate of rotation; but the frequency or wave-length of the light received from a moving source depends upon its velocity and Keeler used this fact to investigate the rotation of the rings. There is little doubt that the fragments of which the ring system is composed are the remnants of a former satellite of Saturn, which approached too close to the planet and paid the penalty, being disrupted by the gravitational pull of Saturn.

The remaining two major planets, Uranus and Neptune, are at such great distances from the Sun that we cannot learn much about them by direct telescopic observation. Uranus, at its mean distance, has an angular diameter of nearly four seconds of arc, which is equal to the angular diameter of a halfpenny placed at a little less than a mile away. The angular diameter of Neptune is about half that of Uranus. Uranus can be seen with the naked eye on a clear dark night, appearing as a star of the sixth magnitude, but Neptune is too faint to be seen without telescopic aid. In the telescope both planets show small disks, which are of a sea-green colour. The disk of Uranus shows a decided oblateness; that of Neptune appears sensibly round. Extremely faint belts, parallel to the equator, have been seen on Uranus by several observers under particularly favourable conditions; they are suggestive of the belts of Saturn seen from a very great distance. There are no markings distinct enough to give a measure of the rotation period, but spectroscopic observations have given a value of above $10\frac{3}{4}$ hours. This is probably a pretty correct value because it has been found that the brightness of Uranus is slightly variable in a period of ten hours forty minutes. The variations in brightness therefore synchronise with the rotation of the planet and suggest that if we could observe Uranus from a closer distance we should see marked differences in the belt structure in different longitudes. Neptune has a slower rotation period of fifteen hours forty minutes; it has the slowest rotation of any of the major planets. This rotation period was measured by the spectroscopic method. Observations of the brightness of Neptune had shown that there was a small regular variation in about half this time—seven hours fifty minutes. So short a period of rotation was difficult to fit in with some theoretical considerations of the orbit of the satellite of Neptune. It is now clear that there must be two portions of the surface of the planet, situated more or less on diametrically opposite sides of the planet, that are brighter than the remainder. From the same theoretical considerations the flattening of the planet can be inferred. It is found to be appreciable, though less than for the other major planets. The little that we can learn by direct observation about Uranus and Neptune suggests, as we should have anticipated, that there is a general similarity between these two distant planets and the two nearer major planets, Jupiter and Saturn. Measurement of the temperature of Uranus indicates that it is below $-180°$ C.; Neptune must be still colder, but its temperature has not been determined by direct observation.

The major planets can be weighed with considerable accuracy because they all have one or more satellites. All that is required for this purpose is to know the period of revolution of a satellite about its parent planet and its distance from the planet; the universal law of gravitation enables us to infer the weight of the planet. Knowing also the size of the planet we can infer its mean density. In all the four major planets the mean density comes out surprisingly low. The mean density of the Earth is $5\frac{1}{2}$ times that of water; the mean density of the

Moon is less, being $3\frac{1}{3}$ times that of water and suggesting that the Moon may have been formed from the less dense outer portions of the Earth. Some people believe that the Pacific Ocean, which is by far the most extensive of the oceans, marks the region where the Moon separated from the Earth. This theory must be looked upon with considerable suspicion, for there are very strong theoretical arguments which suggest that the Earth and the Moon could never have formed one body. The mean densities of Mercury, Venus and Mars all lie between the values for the Earth and the Moon. But when we come to the major planets we find appreciably lower values: the mean densities, in terms of that of water, for Jupiter, Saturn, Uranus and Neptune are respectively 1·34, 0·71, 1·27, 1·58. The low density of Saturn, less than three-quarters the density of water, is particularly surprising; it is only half the mean density of the Sun, which is a gaseous body. The mean densities being so much lower than the densities of any rocks and the planets being known to possess extensive atmospheres, the obvious explanation is that these atmospheres must have very great depth. An appreciable portion of the apparent diameter of these planets is consequently contributed by their deep atmospheres. Our expectation that these massive planets must have extensive atmospheres is thus confirmed.

The temperatures of the giant planets are extremely low. The moisture that must have been present in their atmospheres, when their temperatures were much higher than they are now, must have condensed to form oceans as the planets cooled.

At a later stage, with still further cooling, these oceans froze to form a coating of ice over the surface. It was suggested by Dr. Jeffreys that these planets may be regarded as consisting essentially of a core of rock, generally similar to the Earth in its constitution and of about the same mean density, surrounded by ice-coatings of great depth, above which are very extensive atmospheres. We are unable to penetrate the atmospheres nearly far enough to see these ice-coatings but we can be pretty certain that they exist.

If, then, we picture these planets to be composed of three regions, each sharply differentiated in density—the central rock core, the ice-coating of lower density and the atmosphere of still lower density—it is possible to make approximate estimates of the extent of each region and of the mean densities of the atmospheres. These estimates have to be determined to give the correct mean density and the observed degree of flattening. According to the calculations made by Dr. Rupert Wildt, the rocky core of Jupiter has a radius of about 22,000 miles, so that it occupies only about one-eighth of the whole volume corresponding to the visible disk. The ice-coating is about 16,000 miles in thickness and the depth of the atmosphere is about 6,000 miles. The figures for Saturn are even more astonishing. The rocky core of Saturn is about 14,000 miles in radius; it is covered with a layer of ice some 6,000 miles thick, over which is an atmosphere extending to a height of 16,000 miles. Saturn has the most extensive atmosphere of any of the planets, both absolutely and relative to the size of the planet; this explains why it has the lowest mean density and the most flattened disk of any planet.   It is of interest to note that the total weight of the atmosphere of Saturn is about equal to that of its rocky core.   The data for the more distant planets, Uranus and Neptune, are a little more uncertain; the glacier coating on both of these planets is some 6,000 miles in thickness, whilst the depths of the atmospheres are approximately 3,000 miles for Uranus and 2,000 miles for Neptune. These figures must be regarded as liable to some uncertainty, but there seems to be little doubt that they provide a substantially correct picture of the constitution of the giant planets.

Some interesting consequences follow from the great extent of these atmospheres.   The pressures at the bottom of them are very great.   At the bottom of the atmosphere of Jupiter, for instance, the pressure is fully a million times the pressure at the bottom of the Earth's atmosphere or, in other words, at the surface of the Earth.   At a relatively small depth in the atmospheres, the pressure is great enough to compress each gaseous constituent of the atmosphere to a density nearly equal to that of the corresponding liquid or solid.   It is stated by Wildt that at the bottom of the atmospheres the pressure is great enough to solidify even the permanent gases, including hydrogen and helium.   Little or nothing is known about the properties of substances at these enormous pressures; if we knew more we should probably be in a better position to understand some of the many puzzling phenomena that are presented to us by the atmosphere of Jupiter, phenomena that have been studied in some detail, as briefly described previously.

We have spoken about extensive atmospheres to the giant planets but we have now reached the surprising conclusion that at a relatively small depth in the outer layer of low density the pressure becomes sufficiently great to compress the constituents into the solid or liquid state. The use of the term atmosphere in connection with these planets is, therefore, somewhat misleading, and it would be more correct to speak of the outer layer of low density; this outer layer probably ceases to be gaseous at a depth of not more than a few hundred miles.

The mean densities of the outer layers are low; according to Wildt's calculations they are $0 \cdot 78$ for Jupiter and $0 \cdot 41$ for Saturn. This enables most of the possible constituents to be excluded, for all known gases in the solid or liquid state have densities exceeding $0 \cdot 3$, with the exception of hydrogen and helium. Frozen oxygen, for instance, has a density of $1 \cdot 45$; nitrogen, $1 \cdot 02$; ammonia, $0 \cdot 82$. In addition to helium and hydrogen, the only gases whose densities in the liquid or solid state are lower than the density of the greater portion of the outer layer of Jupiter are the two hydrocarbons, methane or marsh-gas and ethane. There seems to be no escape from the conclusion that the outer layer of the major planets must contain large quantities of liquid or solid hydrogen and helium.

This conclusion is in full accord with expectation. The planets are believed to have been formed in some way from the Sun, so that initially their composition must have been generally similar to that of the outer layers of the Sun. Now we have learnt in recent years a good deal about the composition of the outer layers of the Sun. The principal constituent is hydrogen; the Sun seems, in fact, to consist of hydrogen to the extent of about one-third part by weight. Next in abundance to hydrogen come helium, oxygen and carbon, followed by nitrogen, silicon and the metallic elements. As massive planets, like the four major planets, are able to retain the light constituents of their original atmospheres, it is to be expected that large amounts of hydrogen and helium are contained in their atmospheres at the present time. There is no means, however, by which this conclusion can be tested by observation.

The spectra of the major planets are of great interest and give some partial information about the composition of their outer atmospheres. In the early days of the application of the spectroscope to astronomy, Huggins discovered, by visual examination of the spectrum of Jupiter, that there was a strong absorption band in the orange region of the spectrum and that there were several weaker bands in the green region. These bands appear more strongly in the spectrum of Saturn than in that of Jupiter, but they are not present in the spectrum of the rings of Saturn; this provides conclusive proof that they originate in the atmosphere of Saturn. Uranus and Neptune show for the most part the same bands, but with still greater intensity, together with some additional ones. As we proceed from Jupiter outwards to Neptune we accordingly find that there is a great increase in the selective absorption of light in the yellow, red and infra-red regions of the spectrum. For Uranus and Neptune the absorptions are so strong that most of the yellow and red regions of their spectra are lost by the absorption; it is because of this loss of the light of long wave-length that these two planets appear green when seen in the telescope. More recent investigations by Dr. Slipher, at the Flagstaff Observatory, Arizona, have extended the spectra of the major planets to longer wave-lengths far into the infra-red region and have revealed several more strong absorptions in that region.

When these absorptions in the spectra of the major planets were discovered, and for more than sixty years afterwards, it was not known what their origins were. They had never been observed in any spectra in the laboratory, and for many years they remained one of the unsolved problems of spectroscopy. The clue to their identification first came in the year 1932 from purely theoretical considerations. From the theoretical study of molecular spectra Dr. Wildt came to the conclusion that some of the absorptions agreed in wave-length with bands that should be present in the spectrum of ammonia and that others agreed in wave-length with bands that should be present in the spectrum of methane or marsh-gas, the gas that is given off by decaying vegetation, and is known to the coal-miner as the deadly fire-damp. These conclusions, based on theoretical reasoning, were subsequently confirmed by laboratory observations; the reason why the bands had not been detected in the laboratory before theoretical reasoning had suggested their identification was that a considerable quantity of gas is required for their production with sufficient intensity to be easily observed.

After the origin of the absorptions had been confirmed in the laboratory and the absorptions produced by ammonia and marsh-gas had been fully investigated, the spectra of the major planets were photographed by Dr. Dunham, using the 100-inch telescope at the Mount Wilson Observatory, with optical power far superior to that which had been available to Dr. Slipher. The more detailed investigation that was possible with the aid of this giant telescope showed that there was a complete and detailed coincidence between the spectra of ammonia and marsh-gas obtained in the laboratory and the absorptions shown in the spectra of the major planets.

A comparison between the strength of the absorptions in the spectrum of Jupiter and the strength of the absorptions produced by passing light through a tube of known length containing ammonia gas at atmospheric pressure, enabled Dr. Dunham to conclude that the quantity of ammonia gas producing the absorptions in the spectrum of Jupiter is equivalent to a layer 30 feet thick under standard conditions of temperature and pressure. The amount of ammonia in the atmosphere of Saturn is not so great, the ammonia absorptions being weaker in the spectrum of Saturn than in that of Jupiter. Ammonia has not been detected at all in the spectra of Uranus and Neptune.

In the spectra of all the four major planets the

THE GIANT PLANETS 141

absorptions produced by marsh-gas are much stronger than those produced by ammonia and, as we have already mentioned, Uranus and Neptune appear green because the yellow and red regions of their spectra are to a large extent cut out by the very great intensity of these absorptions. It was found by Drs. Adel and Slipher in 1935, that a column of marsh-gas, forty-five feet in length, and at a pressure of forty atmospheres, gave absorption bands that were intermediate in intensity between those of the bands present in the spectra of Jupiter and Saturn. The much greater strength of the absorptions due to marsh-gas in the spectra of the two more remote planets, Uranus and Neptune, is probably accounted for by the lower temperature of these planets. At the very low temperatures that prevail on Uranus and Neptune, all the ammonia is frozen out of their atmospheres; this explains why no ammonia can be detected in the spectra of Uranus and Neptune. The absence from the atmospheres of these two planets of clouds consisting of droplets of liquid ammonia or small crystals of frozen ammonia, which must be present in the atmospheres of Jupiter and Saturn, makes it possible to see through the atmospheres of Uranus and Neptune to a much greater depth than in the case of Jupiter and Saturn. Adel and Slipher have estimated that twenty-five miles of marsh-gas at atmospheric pressure would be required to give absorptions as strong as those that are present in the spectrum of Neptune. When it is recalled that the equivalent thickness of the atmosphere of the Earth at atmospheric pressure is only $5\frac{1}{2}$ miles, we have here very direct and conclusive evidence of an atmosphere far more extensive than that of our Earth.

It might be expected that, marsh-gas being such an important constituent of the atmospheres of the major planets, other gaseous hydrocarbons would also be present in their atmospheres. Such substances as ethane, ethylene and acetylene have been looked for in vain. Ammonia and marsh-gas between them account, in fact, for all the absorptions detected in the spectra of the major planets and there are no absorptions remaining to be accounted for by other gaseous constituents. It is a grand slam.

The gaseous atmospheres of the major planets, which form the upper regions of their outer low-density layers, are, therefore, entirely different from the atmosphere of the Earth. Hydrogen and helium must be present in these atmospheres in large quantity and the other inert gases, argon, krypton, etc., must also be present. There is a considerable amount of the poisonous marsh-gas and, in the case of Jupiter and Saturn, some of the pungent ammonia gas also; the temperatures are too low for ammonia to be present in large amounts. There cannot be any carbon dioxide, because of the low temperature; there is unlikely to be any free nitrogen and there will certainly be no free oxygen.

How does it came about that we find on these planets atmospheres that contrast so strangely with the atmosphere with which we are familiar on the Earth? Can we explain how such atmospheres have come into existence? In a general way we are able to provide a satisfactory explanation and to show that the surprising difference between the atmospheres of the great planets and the atmosphere of the Earth is the result of the great planets having been able to retain all their hydrogen, whereas most of the hydrogen that was present in the initial atmosphere of the Earth was able to escape whilst the Earth was still hot.

Let us consider what is likely to have been the course of events whilst one of these giant planets cooled. With the gradual fall in temperature from its initial high value, a stage would be reached when liquefaction would set in, giving rise to a liquid core surrounded by an extensive gaseous atmosphere. In the presence of hydrogen at a high temperature, the oxides of iron are reduced, metallic iron being produced. Most of the iron would, therefore, go into the liquid core in its metallic state, oxygen being set free and entering into the atmosphere. The oxides of the remaining principal constituents of the rocks, including potassium, sodium, magnesium, calcium and silicon, are not reduced in the presence of hydrogen at high temperature. They would, therefore, combine to form liquid rock masses, generally similar in composition to terrestrial rocks. With still further cooling, the rock crust would begin to solidify; there would still be a liquid core of molten iron and the atmosphere at this stage would consist of hydrogen, helium, oxygen, nitrogen, carbon and smaller quantities of the inert gases, as well as of sulphur, chlorine and other elements.

As the planet cooled and the temperature fell still further, chemical action would take place. The

PLATE 10

SPECTRA OF THE PLANETS

The plate shows portions of the spectra of Venus, Mars, Jupiter and Saturn, to illustrate the composition of their atmospheres.

The first strip shows corresponding portions of the spectra of the Sun and Venus on the same scale. Many of the absorption lines are common to the two spectra; these absorptions are produced in the outer layers of the Sun. A number of additional strong absorptions are to be seen in the spectrum of Venus, which are not present in the spectrum of the Sun. These are all produced by carbon dioxide in the atmosphere of Venus.

The second strip shows portions of the spectra of Mars (above) and the Moon (below) taken when at the same altitude under conditions of exceptional atmospheric dryness by Dr. Slipher at the Flagstaff Observatory (altitude 7,180 feet). The absorption due to water-vapour (marked a) is stronger in the spectrum of Mars than in that of the Moon, indicating the presence of water-vapour in the atmosphere of Mars.

The third strip shows the spectrum of the ball of Saturn (centre) and of the rings (above and below). Between D and C, and to the right of B, absorptions are seen in the spectrum of Saturn which are absent from the spectrum of the rings. These are caused by marsh-gas in the atmosphere of Saturn.

The bottom strip shows corresponding portions of the spectra of the Sun, Saturn and Jupiter. Absorptions produced by ammonia gas (shown under the spectrum of Jupiter) are present in the spectrum of Jupiter and, less strongly, in the spectrum of Saturn, but are absent from the Sun's spectrum. This proves that ammonia is present in the atmospheres of Jupiter and Saturn.

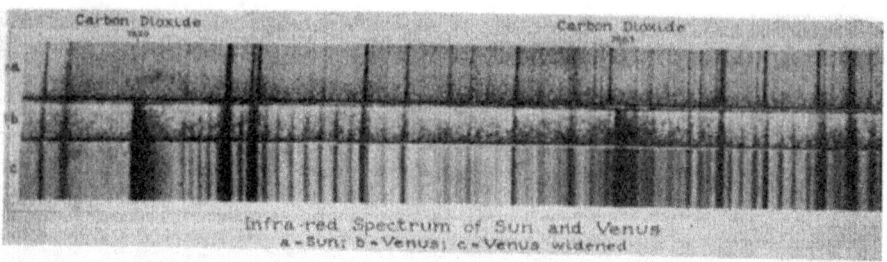

Infra-red Spectrum of Sun and Venus
a = Sun; b = Venus; c = Venus widened

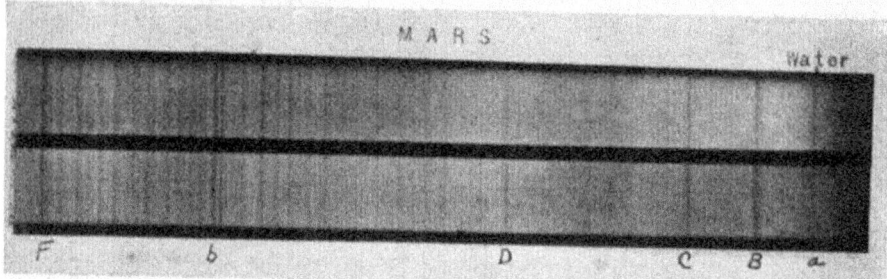

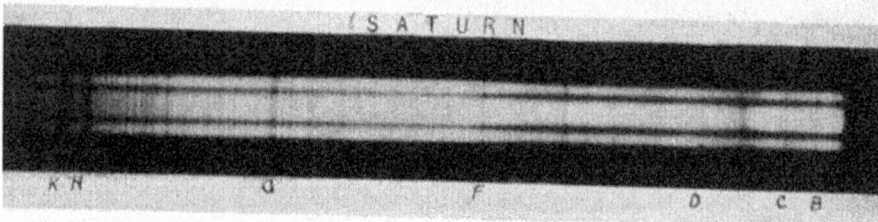

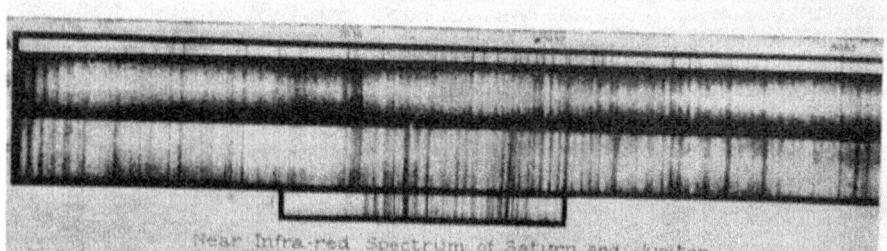

Near Infra-red Spectrum of Saturn and Jupiter
a = Sun; b = Saturn; c = Jupiter; d = Ammonia Gas

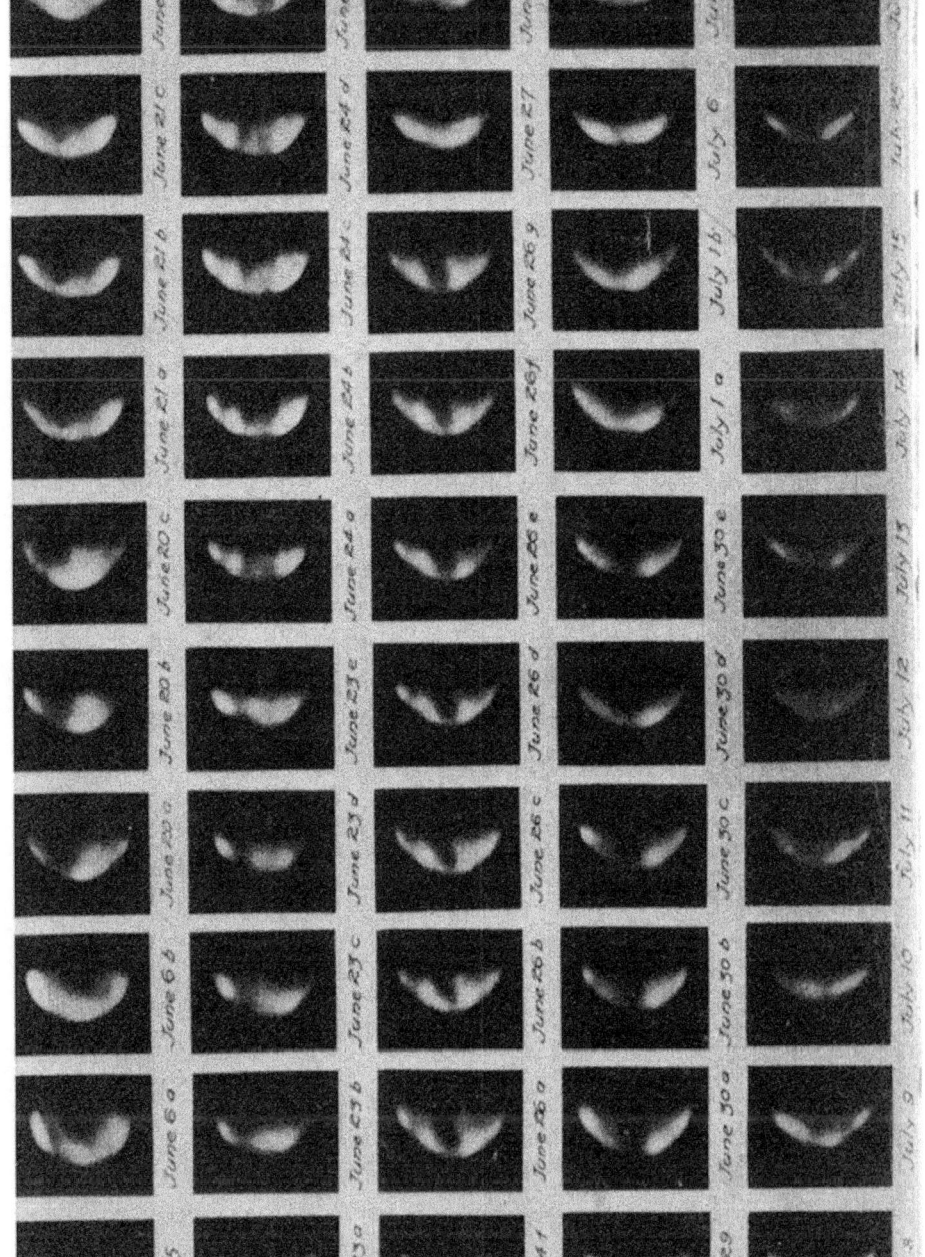

PLATE 11

CLOUDS ON THE PLANET VENUS

A portion of a series of photographs of Venus in ultra-violet light, taken by Dr. F. E. Ross in June 1927 with the 60-inch reflector of the Mount Wilson Observatory, is reproduced. The contrast on the original negatives has been greatly increased by copying in order to make the markings more readily visible. Corresponding markings are not shown on photographs in red light.

A comparison of photographs taken on the same day shows the reality of the recorded markings. These are of two types: bright clouds, which appear as bulges at the limb, and dark clouds, which appear as depressions at the limb. The dark clouds are most frequently seen near the " terminator," the dividing line between the sunlit and dark parts of the planet.

The bright clouds are probably thin cirrus clouds above the permanent cloud layer, overlying a yellowish atmosphere. Where this upper atmosphere is free from cloud, no ultra-violet light is reflected back and the appearance is that of a dark cloud.

carbon would combine with some of the oxygen to form carbon dioxide. The remainder of the oxygen would combine with some of the hydrogen, which, it may be recalled, was by far the predominating constituent of the atmosphere, to form steam. The atmosphere would then consist largely of hydrogen, helium and the other inert gases, carbon dioxide and steam. By this time the temperature would have fallen to somewhere about 1,000° C.

Now carbon dioxide can combine under suitable conditions with hydrogen to form marsh-gas and water-vapour. But the reaction is what chemists term a reversible one. Under other circumstances it can proceed in the reverse direction, so that the marsh-gas then reacts with water-vapour to give carbon dioxide and free hydrogen. The formation of marsh-gas is accompanied by a decrease in volume and is therefore favoured by high pressure, whereas high temperature tends to cause the reaction to proceed in the reverse direction. At the temperature of 1,000° C. and the relatively low pressure which then prevailed on the planet, the predominant tendency is for the mixture of carbon dioxide and hydrogen to be stable; but as the temperature fell there would be an increasing amount of marsh-gas formed. When the temperature had fallen below about 300° C. practically all the carbon dioxide would have been converted into marsh-gas. The presence of marsh-gas or methane in the atmosphere of a giant planet might, therefore, have been anticipated.

We have mentioned that no traces can be detected in the atmospheres of the major planets of other hydrocarbons such as acetylene, ethane, ethylene and so forth. But it is possible for these hydrocarbons also to be formed from carbon dioxide and hydrogen. How, then, are we to account for their absence? The explanation lies in the facts that molecules of these substances are readily broken up by the action of ultra-violet light, and that they are also attacked by hydrogen atoms, the carbon chains being broken up. As a result of such processes the higher hydrocarbons would be completely destroyed in a time that is short compared with the age of the planets, methane being the end result of this process. The molecules of methane are themselves broken up by the action of ultra-violet light, but in an atmosphere containing free hydrogen the methane is again formed. When there is an ample supply of hydrogen, as there certainly is in the atmospheres of the giant planets, a steady state results in which, though molecules of methane are being continually broken up and formed again, there is a constant amount of methane present in the atmosphere.

It has been suggested that the formation of methane and steam from carbon dioxide and hydrogen at a temperature of about 300° C. would be greatly facilitated by suitable activation. The partially reduced oxides of iron, which should be present on the rocky surface of the planet exposed to hot hydrogen, would provide a suitable agent for such activation and would ensure that all the carbon dioxide would rapidly be transformed.

As the temperature fell still further there would be combination between the nitrogen in the atmosphere and some of the hydrogen to produce ammonia. This, again, is a reversible reaction and the ammonia will in turn break down into nitrogen and hydrogen. The lower the temperature and the higher the pressure, the greater will be the concentration of ammonia. It has been shown that the presence of methane will tend to prevent the decomposition of ammonia. The net result is that as the temperature falls below about 200° C. the proportion of free nitrogen to ammonia will decrease rapidly.

The planet continued to cool, and at length the water-vapour began to condense and ultimately an ocean of very great depth was formed. In condensing, the water vapour would carry ammonia gas with it in solution, so that the ocean would be strongly alkaline with ammonia. Compounds of sulphur and chlorine with hydrogen, such as the evil-smelling sulphuretted hydrogen, which may have been present in the atmosphere in small amounts, would also be carried down in solution. With a still further fall in temperature a stage at length arrived when the oceans froze. It may be noted that a solution containing one part of ammonia in two of water freezes at — 100° C.; the major planets are all colder than this.

The final result, after the planet had cooled to about its present temperature, was an atmosphere in which hydrogen is the predominant constituent, with helium present in considerable amount, and argon, neon and the other rare gases present in lesser proportions. There would be no water-vapour, for it had all been frozen out, no carbon dioxide and no

nitrogen. Methane should be present in considerable amount, but ammonia in moderate amount
only, for it had mostly been carried down in solution. In planets as cold as Uranus and Neptune
the residual ammonia would be precipitated as a
solid; even in the atmosphere of Jupiter it is not far
from the point of precipitation.

This fact may be used to obtain an estimate of
the minimum possible temperature of the visible
surface of Jupiter. The amount of ammonia in
the atmosphere of Jupiter is estimated to be equivalent to a layer thirty feet thick under standard conditions. Below a certain temperature it would not
be possible for the quantity of ammonia, which is
observed to be present in the atmosphere of Jupiter,
to exist in the atmosphere; it would be partially precipitated by its own weight. It can be shown from
theoretical considerations that in an atmosphere
consisting mainly of hydrogen, the lowest temperature possible at which the observed quantity of
ammonia could be present is $-120°$ C. This is in
close agreement with the temperature determined
by direct observation. The amount of ammonia in
the atmosphere of Saturn is less than in that of
Jupiter and is consistent with a temperature some
$15°$ C. lower; this, again, is in close accordance
with the observed temperature.

From purely theoretical reasoning, therefore, we
should expect to find atmospheres on the giant
planets which are closely in accordance with what
observation reveals. The crucial fact involved in
their interpretation is the presence of a large excess
of hydrogen; it is to this abundance of hydrogen

that the marked contrast between the atmospheres of these planets and the atmosphere of the Earth is due.   In their broad outlines the main facts about the constitution of these atmospheres are explained. The general line of reasoning will apply to all very massive planets.   We may expect them to have very extensive atmospheres, containing hydrogen, helium and marsh-gas; but no oxygen.

There still remain many points of detail to be explained, such as the nature of the belts on Jupiter and Saturn, the nature of the Red Spot and South Tropical Disturbance on Jupiter and the variety of colours, ranging from white through pink and brown, which are to be seen upon the surface of Jupiter.   It has been suggested that these colours may be due to the formation of small quantities of iridescent compounds involving sodium.   It is known that sodium is present in small quantity in the atmosphere of the Earth at very great heights and the suggestion has been made that the sodium has been swept up by the Earth as it moves with the Sun through space, for sodium is known to be present in very small amount in interstellar space. If this were so—and it must not be regarded as having been definitely established—the presence of sodium in the outer atmosphere of Jupiter could be accounted for in the same way.   This attempt to account for the colours to be seen on Jupiter can scarcely be regarded at present as much more than mere speculation, but it is mentioned here as an indication of the lines along which attempts are being made to explain some of the phenomena that are still very puzzling.   A possible explanation of the Red Spot has recently been suggested by Wildt. He considers that it may be a vast solid body floating in an ocean of permanent gases. The nature of the solid body is not known, but we may think of it as analogous to an iceberg floating in the ocean; it may consist of solid hydrogen. Peek has pointed out that if the level at which it floats is subject to slight variations, the changes in rotational velocity of the spot can be accounted for. In this connection he remarks that whenever the spot becomes conspicuously dark the rotational period lengthens. The changes of colour may be associated with changes of level. It seems that in this way it may be possible to explain the principal phenomena of the Red Spot.

The giant planets are worlds in strange contrast to our own with their enormously deep coatings of ammoniated ice, covered to a depth of thousands of miles with solid or liquid gases, over which are atmospheres devoid of oxygen or water-vapour but containing large quantities of poisonous marsh-gas. These dreary, remote, frozen wastes of the solar system are not worlds where we can hope to find life of any sort. Great cold may not by itself make life impossible, even though it may make its development extremely improbable; nor by itself may great pressure. But when these conditions are combined with absence of oxygen and of moisture and with abundance of poisonous gases, we have such a combination of unfavourable circumstances that we are compelled to turn elsewhere in our quest for life in the universe.

CHAPTER VII

VENUS—THE EARTH'S TWIN SISTER

OF all the planets in the solar system, Venus most nearly resembles the Earth in size, in mass and in mean density: and so it is on Venus that we have the greatest expectation of finding conditions akin to those that exist on the Earth.

The path of Venus lies inside the path of the Earth. When Venus is at its nearest to the Earth it is only 26 million miles away. No other body ever comes so near the Earth, with the exception of the Moon and an occasional comet or asteroid. When at its farthest from the Earth, Venus is 160 million miles away. With such a wide range between its greatest and least distances, it is natural that at some times Venus appears much brighter than at others. When at its brightest it is easily seen with the naked eye in broad daylight. When I lived in South Africa, I was sitting by the sea one day when, by chance looking upwards, I saw, as I thought, an aeroplane flying at a great height, glittering brightly in the sunshine against the azure blue sky. But as I watched, I noticed that it remained in the same position, and only then realised that I was looking, not at an aeroplane, but at the planet Venus.

When the orbit of a planet lies outside the orbit of the Earth the planet is, at times, visible throughout the night. When, however, the orbit is within the orbit of the Earth, the planet can never be seen at midnight. It is visible either in the evening after sunset or in the morning before sunrise. Venus, because of its brightness, becomes a conspicuous object in the twilight sky and is often referred to as the evening star or the morning star according as it is seen after dusk or before dawn. The Greeks had two names for it; they called it Phosphorus when it appeared at dawn and Hesperus when it appeared at dusk. The identity of the evening and morning star was known, nevertheless, as early as the time of Pythagoras.

Suppose we observe Venus night after night with a telescope. Commencing when it first appears low in the western sky just after sunset, we find that it shows a full sun-illuminated disk; as it draws gradually away from the Sun night after night for nearly eight months it appears progressively brighter and larger, but the portion of the disk that is visible becomes gradually less. When at its greatest elongation from the Sun, it shows the half-moon phase. Then it will be seen to turn back, drawing gradually nearer to the Sun and developing a crescent phase that becomes narrower and narrower; the diameter of the image continues to increase; and for a time the planet continues to become brighter. The greatest brightness is reached about 32 days after the greatest eastern elongation, the phase then corresponding to that of the Moon when five days old. About ten weeks after eastern elongation, Venus vanishes in the Sun's rays; greatly diminished in brightness, it then shows a large but narrow crescent. A little later it appears as the morning star and corresponding changes are passed through in the reverse order.

It is of interest to recall that the phases of Venus were first discovered in 1610 by Galileo, who was able to see the changing phases in his telescope. As was frequently done at that time, he announced his discovery in the form of an anagram:

Haec immatura a me iam frustra leguntur : o.y.

The solution of the anagram, which concealed his discovery, was given some months later as:

Cynthiæ figuras æmulatur Mater Amorum,

which means that " The Mother of the Loves (i.e. Venus) imitates the phases of Cynthia (i.e. the Moon)."

Venus has been called the twin sister of the Earth because it closely resembles the Earth in size and weight. It is a little smaller than the Earth, its diameter being 7,700 miles as compared with the Earth's 7,927 miles. The area of its surface is therefore five per cent. smaller than the area of the surface of the Earth. Its weight is about four-fifths of the weight of the Earth. The velocity of escape from Venus is only slightly less than the velocity of escape from the Earth and it is therefore to be anticipated that Venus will be found to possess a fairly extensive atmosphere, comparable in extent with the atmosphere of the Earth.

The presence of an atmosphere on Venus is readily proved by observation. When Venus is between us and the Sun, showing the narrow crescent phase, the tips of the horns of the crescent are not at the two ends of a diameter as they are in the case of the crescent Moon, but the horns extend well round the circumference of the dark limb. This means that there is a region of twilight on Venus; the Sun is not shining directly on this portion of the planet but it becomes faintly visible by the light scattered in the atmosphere. If the Earth had no atmosphere, darkness would come suddenly as soon as the Sun had set; the scattering of the sunlight by the Earth's atmosphere for some time after the Sun has disappeared below the horizon causes the gradual transition from day to night, which we term twilight.

Direct evidence of an atmosphere on Venus is also obtained at times of the transits of Venus. I have mentioned that the path of Venus lies inside the path of the Earth and it must sometimes happen that Venus will pass directly between us and the Sun. We then see Venus as a dark spot moving across the face of the Sun. If the orbits of Venus and the Earth were in the same plane, Venus would pass in front of the Sun every time that it changed from an evening to a morning star. The orbit of Venus is, however, inclined to the orbit of the Earth as an angle of about $3\frac{1}{2}°$; Venus can therefore only be seen projected on the disk of the Sun at a time when it is close to one of the two points where its orbit crosses the plane of the orbit of the Earth. Transits of Venus across the Sun are somewhat rare phenomena. The last two transits occurred on December 8th, 1874, and December 6th, 1882; the next two transits will occur on June 7th, 2004 and June 5th, 2012. When

a transit of Venus takes place and Venus is just
entering upon the disk of the Sun or just leaving it
the edge of the portion of Venus that is outside the
Sun is surrounded by a bright line of light. This
appearance can only be caused by the scattering of
sunlight by an atmosphere around Venus. It will
be remembered that this appearance is not seen
when Mercury transits on to the Sun's disk, because
Mercury has no atmosphere.

When we look at Venus in the telescope, we are
apt to be disappointed. As we have already
mentioned, she shows phases like the Moon,
depending upon the proportion of her sunlit face that
is turned towards us. We might hope to find some
evidence of continents and oceans on a world that
is in some respects so similar to our own. But we
are doomed to disappointment. There are no
well-defined markings to be seen on her surface.
Nothing more than faint ill-defined shadings may
be seen, and these can only be seen occasionally.
They are very elusive, for there is so little contrast
between the markings and the rest of the disk that
they are barely visible. They have been described
as " large dusky spots " and are certainly not
permanent. Such markings as are from time to
time visible are therefore certainly not surface
markings; they must be produced in the planet's
atmosphere.

Venus must either be covered with a permanent
layer of cloud or her atmosphere must be so hazy
that light from the Sun cannot penetrate to the
surface and out again. The light that reaches us
from Venus must either have been reflected from

a layer of cloud or have been scattered within the atmosphere without penetrating to the surface.

It might be hoped that some further information could be obtained by photographing Venus on plates sensitive to the long wave-length infra-red light. Within recent years there has been a great development in photography in the use of infra-red or haze-cutting plates. Everybody is familiar with the way in which, on a fine day, with a slight haze, the detail in a distant landscape is lost. But if, under such conditions, we take a photograph using a plate that has been specially sensitised for the infra-red light, much detail that is invisible to the eye will be clearly shown. This is illustrated in Plate 12 where two photographs are reproduced, showing the view from the top of Mount Hamilton in California, where the Lick Observatory is located, looking across the valley to the distant mountains. One photograph was taken with a plate sensitive to the short-wave ultra-violet light, the other with a plate sensitive to the long-wave infra-red light, the two photographs being taken at the same time and therefore under the same conditions. The ultra-violet photograph dimly shows the skyline of the distant mountains but none of the intervening detail, which is clearly revealed by the infra-red photograph.

We therefore photograph Venus using these infra-red plates, in the hope that in this way something will be revealed that the eye cannot detect. Venus, however, will not be circumvented in this way. She refuses to reveal her secrets; the plate shows us no more than our eyes can see. The infra-red light is no more successful in penetrating to the surface and out again than the ultra-violet light had been.

There is one difference that may be noted between the photographs in ultra-violet light and those in infra-red light. The former record bright markings, which rapidly change their form and are of short duration. Photographs of Venus in ultra-violet light, showing some of these bright markings and illustrating the rapidity with which they change, are reproduced in Plate 11. These markings are not seen on the photographs in the infra-red light. They indicate the existence of some sort of atmospheric haze or possibly of very thin clouds at a high altitude, through which the light of long wave-length can pass without difficulty, but which scatters or reflects the light of short wave-length.

Venus reflects about 60 per cent. of the sunlight that falls upon it. This high reflecting power, which contrasts with the low reflecting powers—about seven per cent.—of both Mercury and the Moon, is very much what we should expect from a planet covered with a thick layer of clouds. The visible surface is practically white—almost devoid of colour; this again is consistent with the appearance of a cloud layer. An estimation of the height to which the atmosphere of Venus extends can be made from the amount of the prolongation of the horns, when the planet is seen as a narrow crescent. It can be shown that the portion of the atmosphere where the twilight is sufficiently bright to be seen through the glare of our own atmosphere extends to a height of about 4,000 feet above the visible surface of Venus. The total height of the atmosphere of Venus must, of course, be many times greater than this, for the faint twilight effect produced by the more tenuous upper reaches of the planet's atmosphere would be lost in the glare of our own atmosphere. It seems probable, nevertheless, that the atmosphere of Venus above the visible surface is both less extensive and less dense than the atmosphere of the Earth. We can more fairly compare it with the portion of the Earth's atmosphere above the high clouds. This seems to suggest that these observations by no means reveal the full extent of the atmosphere of Venus, which we should expect to be fairly comparable with that of the Earth. It therefore seems reasonable to conclude that the visible surface is a permanent layer of cloud, which we have no means of penetrating, situated at a fairly high level above the surface of the planet.

The ill-defined nature of the markings seen on Venus and their impermanence make it difficult to determine the length of the day on Venus with any accuracy. The conclusions reached by different observers have been very varied and somewhat contradictory. Some have asserted that the length of the day on Venus is about the same as the length of the day on the Earth; others have concluded that Venus always turns the same face to the Sun, just as Mercury does, so that her day is equal to her year, which is 225 of our days. Most probably the truth lies between these two extremes. It was through the agency of the tides raised by the

PLATE 12

PHOTOGRAPHS OF MARS AND OF TERRESTRIAL LANDSCAPE

The plate shows photographs of Mars (*a* and *b*) and of San José, taken from the top of Mount Hamilton, California (*c* and *d*). The photographs *a* and *c* were taken with ultra-violet light (short wave-length), *b* and *d* were taken with infra-red light (long wave-length). The distant mountains and the town of San José in the valley are closely seen in *d* but are obliterated in *c*. The distance of San José from the point where the photograph was taken is 13½ miles; the short wave-length ultra-violet light was unable to penetrate through this extent of the Earth's atmosphere. The surface markings of Mars are clearly shown in *b*, but not in *a*, indicating the presence on Mars of an atmosphere of sufficient extent to prevent the ultra-violet light from penetrating to the surface of Mars and out again.

Photographs by Dr. W. H. Wright, Lick Observatory, with Crossley Reflector.

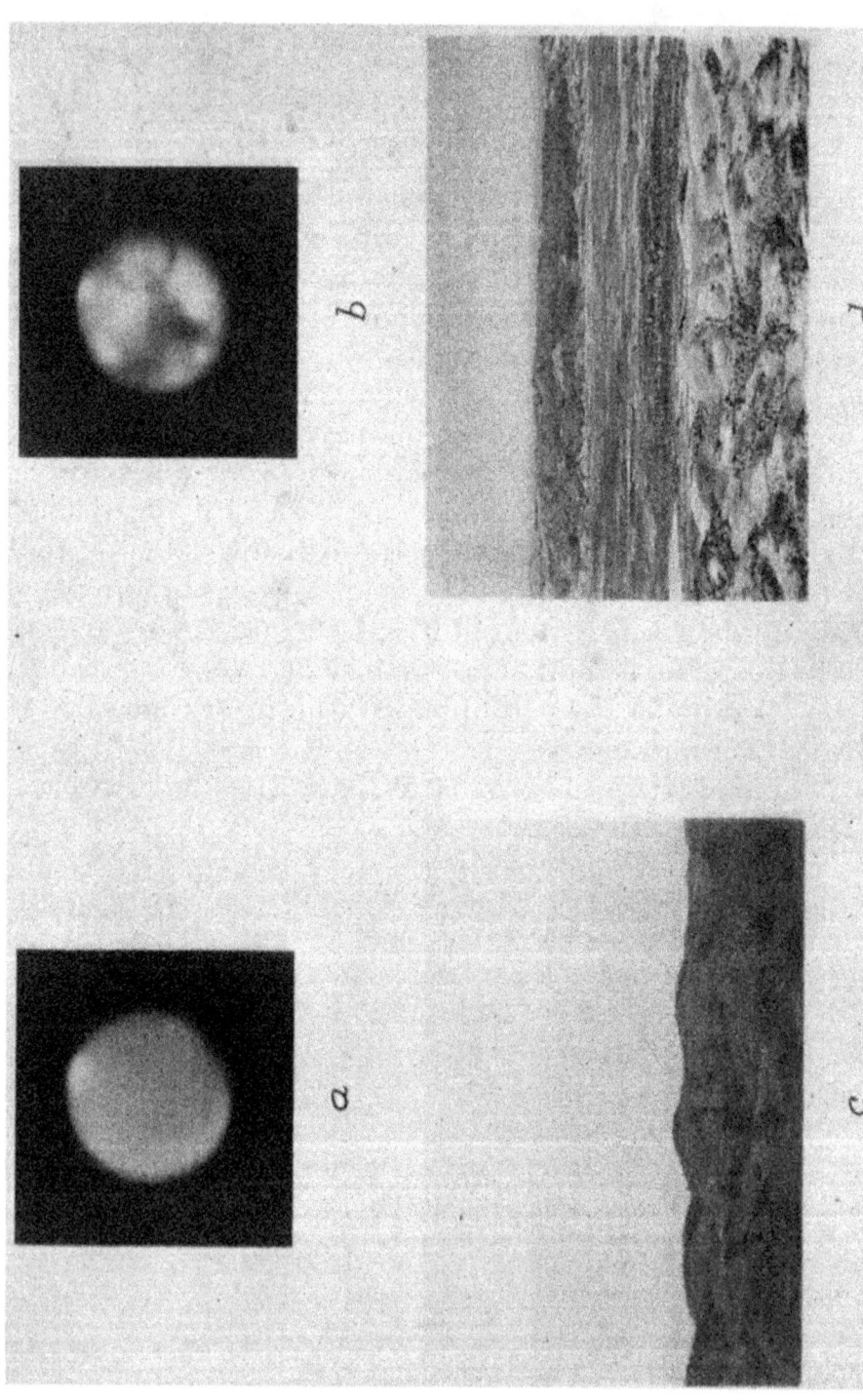

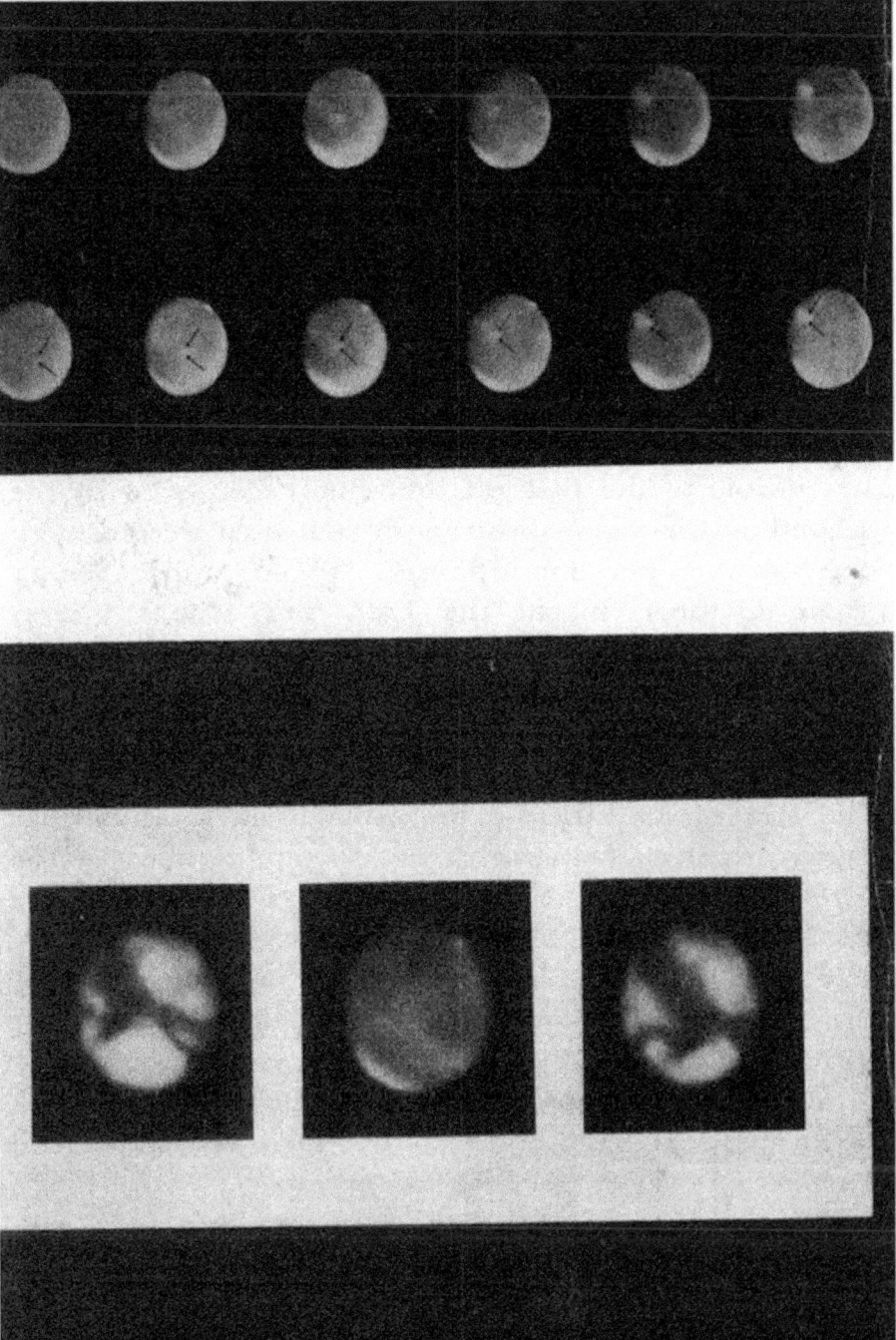

PLATE 13

CLOUDS ON THE PLANET MARS

The series of six photographs, taken in ultra-violet light, show the formation and growth of a white cloud during the course of a Martian afternoon. The photographs, in order from left to right, were secured at intervals through a period of about four hours, during which the planet rotated 55°. The place of the brightest part of the cloud, in the successive positions of the planet, is shown by the arrows in the lower row, the rotational movement being upwards and to the left. The cloud is not visible in the first picture, but becomes so in the second and increases in strength as it is carried through the Martian afternoon to sunset. Photographs taken on 1926, October 16, at the Lick and Mount Wilson Observatories.

The lower series of photographs shows the formation of a yellow cloud, visible in infra-red but not in ultra-violet light. The centre photograph is taken in ultra-violet light, the others in infra-red light, the markings being displaced between these two photographs by the rotation of Mars. If the photographs are turned 90°, the markings will be seen to have the appearance of a stag's head. In the left-hand photograph will be seen, just under the stag's neck, a bright patch, which is not visible on the right-hand photograph, nor in the photograph in ultra-violet light. This bright patch is a yellow cloud. Photographs by Dr. Wright, 1926, November 2–3, at the Lick Observatory.

action of the Sun on Mercury before it solidified that the rotation of Mercury was slowed down until it eventually turned the same face always to the Sun; similarly, the tides raised on the Moon by the attraction of the Earth slowed down the rotation of the Moon until it always turns the same face to the Earth. The tides raised on Venus by the gravitational attraction of the Sun would have been much smaller than those raised on Mercury, because Venus is at a considerably greater distance. The tide-raising force falls off inversely as the cube of the distance, and at the distance of Venus it is only about one-sixth as great as at the distance of Mercury. It is consequently to be anticipated that, though the rotation of Venus would be slowed down to a moderate extent by the friction of the tides raised on her by the Sun, the effect would not be nearly large enough to cause her always to turn the same face to the sun.

That the rotation period of Venus is not as short as a day is quite certain. If her rotation were as rapid as this, there would be no difficulty in detecting it by comparing the spectra of the light from the east and west limbs. The relative motion of the two limbs, one moving towards the Earth and the other away, would produce small relative shifts between the wave-lengths of the corresponding lines in the two spectra, which would be readily observed. No such shift has been detected. The inference from this failure to detect a relative shift is that the period of rotation must be at least several weeks in length. The same conclusion has been reached recently by M. Antoniadi from observations of the faint markings seen on Venus with the aid of the great telescope at Meudon, near Paris.

It seems reasonably certain, on the other hand, that Venus does not turn the same face always to the Sun. The temperature of the sunlit side of Venus has been found by observation to be about 50° to 60° C., whilst that of the dark side is about — 20° C. A much lower temperature of the dark side would be expected if it never received any heat from the Sun, but received heat only through convective motions of the atmosphere. The sunlit face, on the other hand, would be very much hotter than it is observed to be if it were continuously receiving heat from the Sun. The difference between the day and night temperatures on Venus is, therefore, considerably smaller than it would be if Venus always turned the same face to the Sun. We shall probably not be far wrong if we assume that the length of the day on Venus is equal to about four or five of our weeks. The year on Venus being equal to 225 of our days, it follows that on Venus there are only some six or seven days in the year.

The measurements of the temperature on Venus do not refer to the true surface of the planet nor to what we may term the visible surface. The temperature of the true surface of the planet, beneath the permanent layer of cloud, is almost certainly appreciably higher than the temperature found by observation. There must be a considerable green-house effect beneath the cloud layer, the short-wave radiation from the Sun being absorbed and given out again as long wave-length heat radiation. It is quite possible that at the surface of

Venus, in the equatorial regions, the temperature may be as high as, or even higher than, that of boiling water.

The light reflected from Venus to the Earth has been analysed to find out whether it shows the absorptions that are characteristic of oxygen or water-vapour. Neither oxygen nor water-vapour has been detected. This negative result does not necessarily imply that there is neither oxygen nor water-vapour in the atmosphere of Venus, but merely that the amount is not sufficient to be revealed by the tests that can be used. The tests for one substance may be extremely sensitive and the presence of a very small quantity can then be detected, whilst the tests for another substance may be so insensitive that it must be present in very great abundance before there can be any hope of detecting it. The most striking illustration of differences in the sensitivity of the tests for the presence of different substances is provided by the comparison between calcium-vapour and hydrogen in the outer layers of the Sun. The calcium-vapour gives rise to two extremely strong absorptions in the spectrum of sunlight, absorptions which are by far the strongest in the whole range of the Sun's spectrum that we are able to study. These intensely strong absorptions are produced by an amount of calcium-vapour which, under standard conditions in the laboratory, would be less than half an inch in thickness. The absorptions produced by hydrogen are weak relatively to those produced by calcium, yet the hydrogen is so abundant that it is estimated that, atom for atom, hydrogen is at least 300 times as abundant as the whole of the metallic vapours together.

The tests for the presence of water-vapour in the atmosphere of Venus are less sensitive than those for the presence of oxygen. An amount of oxygen on Venus equal to a thousandth part of that above an equal area of the earth could certainly have been detected. It must be remembered, however, that we are unable to see down to the surface of Venus and any tests that we can apply give information only about the portion of the atmosphere that is above the permanent layer of cloud. Nevertheless, if the oxygen above the cloud layer were as much as the one-hundredth part of the oxygen in the atmosphere of the Earth above the highest clouds, it would have been detected. There is no escape from the conclusion that there can be very little, if any, oxygen in the atmosphere of Venus.

The failure to detect water-vapour, even though the tests are less delicate than those for oxygen, may appear at first sight to be rather surprising. For if Venus is covered with a permanent layer of clouds there must be water-vapour, unless the clouds are of a different nature from those that occur in our own atmosphere. No alternative suggestion, which is at all feasible, of the nature of the clouds on Venus has been made and it seems certain that they must consist of water droplets, similar to the clouds in the Earth's atmosphere. The explanation of the apparent absence of water-vapour must be that the atmosphere above the clouds is extremely dry. Most of the water-vapour has been condensed out of the upper atmosphere and what is left is not sufficient for us to be able to detect. In this respect the atmosphere of Venus seems to be similar to that of the Earth. In the atmosphere of the Earth the water-vapour is confined almost wholly to the lower layers and the amount above a height of five miles is always small. If the cloud layer on Venus reaches to a height of four or five miles above the surface, the failure to detect water-vapour in its atmosphere presents no difficulties.

The most interesting and significant fact about the atmosphere of Venus is the great abundance of carbon dioxide. In 1932, Adams and Dunham, using the great 100-inch telescope at the Mount Wilson Observatory, discovered three strong absorptions in the spectrum of Venus in the long wavelength infra-red region. These absorptions are not found in the spectrum of the Sun, even when setting. They are not produced, therefore, in the atmosphere of the Earth but must originate in the atmosphere of Venus. At the time that they were discovered they had not been observed in any terrestrial spectrum and it was not known, therefore, what substance produced them. Theoretical investigations suggested that they might be due to carbon dioxide. This was confirmed when Dunham succeeded in obtaining a faint absorption, corresponding in position with the strongest of the three absorptions, by passing light through forty metres of carbon dioxide at a pressure of ten atmospheres. Later, Adel and Slipher were able to obtain all three absorptions by passing light through forty-five metres of carbon dioxide at a pressure of forty-seven atmospheres. The three absorptions coincided exactly in position with those observed in the spectrum of Venus but, even with so great a thickness of carbon dioxide, they were less intense than the corresponding absorptions in the spectrum of Venus.

Their laboratory investigations enabled Adel and Slipher to conclude that the amount of carbon dioxide above the visible surface of Venus is equivalent to a layer two miles in thickness at standard atmospheric pressure and temperature. This affords direct experimental confirmation of the conclusion, which we have already reached, that Venus has an extensive atmosphere. The interest of the result lies, however, in the comparison between the amount of carbon dioxide on Venus and the amount in the Earth's atmosphere. The amount of carbon dioxide present in the path of sunlight, when the Sun is setting, is equivalent to a thickness of about thirty feet only. Carbon dioxide is accordingly vastly more abundant in the atmosphere of Venus than in that of the Earth.

The amount of carbon dioxide observed to be present in the atmosphere of Venus, equivalent to a layer two miles in thickness at standard atmospheric pressure and temperature, represents only the portion of the carbon dioxide that is above the permanent cloud layer—the visible surface. The total amount above the solid surface of the planet may well be appreciably greater. When we recall that the whole atmosphere of the Earth at standard pressure and temperature would form a layer only about $5\frac{1}{2}$ miles in thickness and that we should expect the atmosphere of Venus to be rather less extensive than that of the Earth, we are forced to the conclusion that carbon dioxide is an important constituent in the atmosphere of Venus, and that it may well be the predominant constituent.

In Chapter IV we considered the evolution of the Earth's atmosphere in some detail. We came to the conclusion that most of the atmosphere that the Earth originally possessed was lost very early, whilst the Earth was still molten. As the Earth cooled and solidification set in, large quantities of gases, principally water-vapour and carbon dioxide, must have been evolved from the semi-molten mass. These gases with the residual gases from the initial atmosphere, consisting mainly of nitrogen, argon, neon and probably some carbon dioxide, formed the new atmosphere.

The same general course of evolution is likely to have occurred in the case of Venus, which is so nearly equal to the Earth in size and weight. The atmosphere of Venus, in the early stages after solidification set in, consisted of carbon dioxide and water-vapour, along with nitrogen, argon, neon and probably small but relatively unimportant quantities of other gases. Nitrogen, being a left-over from the initial atmosphere, is probably present in the atmosphere of Venus in smaller amount than in the atmosphere of the Earth, because it would be somewhat easier to escape from Venus than from the Earth.

When Venus cooled to a temperature below the boiling-point of water, most of the water-vapour would condense out of the atmosphere and form oceans and lakes, leaving an atmosphere consisting primarily of carbon dioxide and nitrogen—the quantity of nitrogen being less, however, than that in the Earth's atmosphere—with some argon and small quantities of neon and other gases.

The Earth at one stage in its history had an atmosphere of a similar nature, but we have seen that the atmosphere of the Earth then passed through a further stage, which consisted essentially in the removal of the carbon dioxide and its replacement by oxygen. It is clear that the atmosphere of Venus has not passed through this stage, at any rate to any great extent. The abundance of carbon dioxide in its atmosphere and the scarcity, or possibly even the absence, of oxygen are both confirmatory.

This is a very important and significant conclusion. What does it tell us? When considering the atmosphere of the Earth, we remarked that the presence of free oxygen demanded explanation because oxygen is chemically a very active substance and there are agencies in continual action tending to deplete the supply. The only way in which it was possible to account for the presence of free oxygen in the atmosphere of the Earth was through the action of vegetation, and it appeared that the store of oxygen in our atmosphere has been brought about through vegetation being buried and thereby preserved from decay—vegetation that is now represented by the coal and oil beneath the surface of the Earth. If there were vegetation on Venus to any great extent, the atmosphere of Venus would also have passed through the corresponding stage of evolution. It clearly has not done so.

Venus is, in fact, very much like what the Earth was before life had commenced to develop.

We can draw a picture of Venus that is probably pretty near the truth, despite the fact that we have never seen her surface. An atmosphere rich in carbon dioxide will have a very powerful blanketing effect. This effect, together with solar radiation stronger than the Earth receives, will combine to make the temperature at the surface of Venus considerably higher than the temperature of the Earth. It is likely that the temperature is not much below 100° C., the temperature at which water boils. Under such circumstances it is most improbable that life would begin to develop. The lack of vegetation and the absence of oxygen from the atmosphere of Venus are thereby explained. There are doubtless extensive oceans and swamps and a very hot humid atmosphere; the abundance of moisture is responsible for the permanent layer of thick clouds.

Venus, then, appears to be a world where life has not yet developed, or, if it has commenced, where it is merely in such a primitive stage that we cannot obtain any direct evidence of it. It is a world where conditions are not greatly different from those that existed on the Earth many hundreds of millions of years ago. There may be expectations of life in the remote future when, as the Sun's supply of radiation becomes gradually depleted and the Sun slowly cools down, conditions will approximate more and more to those that the Earth passed through and which led eventually to the appearance of life. As conditions become more suitable for life to appear on Venus they will become less favourable for its continued existence on the Earth. After life on the Earth has become extinct, a new chapter may commence on Venus leading gradually and progressively to more and more highly developed forms of life and ultimately—who can tell ?—to intelligent life.

CHAPTER VIII

MARS—THE PLANET OF SPENT LIFE

To many persons Mars is the most interesting object in the heavens because it is the one and only world where we appear to have direct evidence of life and because some astronomers have held the opinion that it provides evidence for the existence on it of intelligent beings.

Mars revolves in an orbit that is outside the orbit of the Earth, its mean distance from the Sun being a little more than one and a half times that of the Earth. The orbit is rather elliptical and, in consequence, the distance of Mars from the Sun varies by more than 26 million miles. It requires a period of a little short of two years for Mars to complete one revolution in its orbit, and we overtake it on an average once in every two years and fifty days. Mars is then said to be in opposition, because the Sun, the Earth and Mars are nearly in a straight line (they would be exactly in a straight line if it were not for the fact that the orbits of Mars and of the Earth are not quite in the same plane), the Sun and Mars being on opposite sides of the Earth, so that Mars rises at sunset and crosses the meridian at midnight. Because the orbits of Mars and the Earth are both somewhat elliptical the distance of Mars from the Earth at opposition can range from about 35 million miles to about 63 million miles. The nearer Mars is to the Earth at opposition, the more favourable the opportunity for studying its surface. The most favourable oppositions occur in August and the least favourable in February. The planet is at its brightest at opposition. At the least favourable oppositions Mars is not quite as bright as Sirius, the brightest star in the sky, whilst at the most favourable it becomes much brighter than any star and brighter than any other planet at its brightest except Venus. When at its greatest distance from the Earth, Mars is about half as bright again as the pole star.

The diameter of Mars is about 4,215 miles, only a little more than half that of the Earth. Its weight is rather more than one-tenth of the weight of the Earth. The force of gravity on its surface is only two-fifths as great as that on the surface of the Earth and the velocity of escape from Mars is about three miles a second. This is less than one-half of the velocity of escape from the Earth, and we should therefore expect to find that Mars has some atmosphere, but that this atmosphere will prove to be considerably more tenuous and less extensive than the atmosphere of the Earth.

The opportunities for the satisfactory observation of Mars are somewhat limited. Its apparent diameter ranges from $3\frac{1}{2}$ seconds of arc, when at its greatest distance, to twenty-five seconds at the most favourable opposition. At the favourable oppositions the size of the image seen in a telescope is therefore about seven times greater, and the area of the image about fifty times greater than when the planet is at its greatest distance. For the study of fine detail on the surface of the planet, the conditions are really only satisfactory for a few months around opposition or, in other words, for a few months every two years or so.

Suppose we have at our command a large telescope of twenty-five feet focal length. At the most favourable oppositions, the diameter of the image of Mars in the focal plane of the telescope is only $\frac{1}{28}$ inch; at the least favourable oppositions, it is about half as large whilst, when at its greatest distance, the diameter of the image is only $\frac{1}{200}$ inch. The smallness of the image in even a large telescope makes it impossible to study in minute detail the surface markings on Mars by photography. The detail is so intricate that much of it is finer than the grain of the photographic plate; moreover, the planet is never bright enough to be photographed by an instantaneous exposure. A time exposure is needed and then the slight tremors in the atmosphere, which almost invariably are present in greater or less degree, blur out the finer details in the image. If we endeavour to reduce the trouble arising from the coarseness of the grain of a fast plate by using slow fine-grained plates, it becomes necessary to increase the time of exposure considerably and the troubles from atmospheric unsteadiness then become greater. So in either case there is a limit to what the photograph can reveal. That is the reason why photographs of Mars show less detail than is recorded in drawings by competent observers. In visual observations, it is possible to wait for the occasional instants when the atmosphere becomes momentarily steady and the detail sharply defined, for on most nights there occur brief moments when the seeing conditions become much better than the average.

When Mars is at its nearest to the Earth, it is more favourably placed for observation than any other heavenly body, with the exception of the Moon. It is true that Venus at times comes closer to the Earth than Mars ever does but, at such times, Venus appears as a very narrow crescent and is visible only by daylight. If, when Mars is most favourably placed, we observe it in the telescope with a power of seventy-five, it appears as large as the Moon does to the naked eye. It might seem at first sight that the detailed study of the surface under such conditions would be easy. But if we compare the coarse detail that we can see on the Moon with the naked eye with the intricate fine detail shown on a photograph obtained with a large telescope, it will be realised how much is lost.

Mars exhibits slight phases in the telescope, which were discovered by Galileo in 1610. But, since its orbit lies outside the orbit of the earth, it can never appear as a crescent, like Mercury and Venus. The greatest phase shown by Mars is comparable with the phase of the Moon when it is three days from full.

Under favourable conditions Mars appears in the telescope as a beautiful object, with a strong orange colour, on which misty markings can be seen. The first sketch showing surface markings on Mars was made in 1659 by Huyghens, who, by studying the apparent movements of these markings, suggested that Mars rotated in twenty-four hours. In 1666 Cassini found the period of rotation to be twenty-four hours forty minutes; this is close to the period given by modern observations, which is about twenty-four hours thirty-seven and a half minutes. Cassini also observed the characteristic feature of the polar caps, but it was not until near the end of the eighteenth century that Sir William Herschel detected the variation of the size of the polar caps with the seasons.

Around whichever of the poles of Mars happens to be visible there is seen a bright white cap. The two polar caps show regular seasonal changes in size. During the northern summer, for instance, the northern cap shrinks whilst the southern cap grows. With the changing of the seasons the northern cap grows again whilst the southern cap shrinks. Analogy strongly suggests something similar to the regions of ice and snow around the north and south poles of the Earth.

In contrast to the changes in the polar caps, which are easily seen in moderate-sized telescopes, the dark markings are more or less permanent. We see these markings carried round by the rotation of Mars, and this enables the period of rotation to be fixed with high accuracy.

The first really detailed and careful survey of the surface of Mars was made by the Italian astronomer, Schiaparelli, at the favourable opposition of 1877. Schiaparelli was a highly competent observer, he had an excellent telescope, good conditions for observing, and Mars was unusually close to the Earth. The existence of dusky markings on the planet, which stood out against the ruddy background, was already known. It was believed that these dusky markings were seas and that the ruddy background was dry land. But in 1877 Schiaparelli discovered, what had not previously been known, that there were dusky streaks crossing the land areas or " continents," connecting up the " seas " with one another. He termed these streaks *canali* which, interpreted literally, means channels. But the similarity of the Italian word to the English word canal has caused a narrower interpretation to be placed upon the term given by Schiaparelli than he intended, with the result that there has been a good deal of misrepresentation.

Schiaparelli continued to observe Mars for a number of years and discovered that the seas were not uniformly dusky. He remarked that the colour of the seas was generally brown, mixed with grey, but not always of equal intensity in all places, nor always the same in the same place at all times. From an absolute black it might descend to a light grey or to an ash colour. He compared these differences in colouration on Mars with differences in colour of the seas on the Earth, adding that the seas of the warm zone are usually much darker than those near the pole; the Baltic, for instance, has a light muddy colour that is not observed in the Mediterranean. Schiaparelli noticed that some at least of the changes in colouration were seasonal in nature; he interpreted these changes as being changes in colour of the seas, which became darker as the Sun approached their zenith and summer began to rule.

It was noticed also by Schiaparelli that the so-called continents were not uniform in colour. Over

12

the bulk of the continents an orange colour pre-
dominated which in some areas, of relatively small
extent, reached a dark red tint, whilst other small
regions were yellow or white.   But besides the dark
and light regions, which were considered to be seas
and continents respectively, there were some areas
of small extent that were of an amphibious nature,
appearing sometimes yellowish like the continents,
sometimes brown or even black like the seas,
whilst in other cases the colour was intermediate in
tint, leaving a doubt whether they were sea or land
areas.   He concluded that these represented huge
swamps, in which the variation in the depth of the
water produced the diversity of colour.

The vast extent of the continents was furrowed
upon every side by a network of numerous lines or
fine stripes of a more or less pronounced dark
colour, whose aspect was very variable.   They
traversed the planet for long distances in regular
lines, not at all resembling the winding courses of
rivers on the Earth.   Some of the shorter ones
were only a few hundred miles in length, but others
extended for thousands of miles, extending over as
much as one-third of a circumference of Mars.
Some of these lines or channels (*canali*) were very
easy to see; others were extremely difficult and
resembled the finest thread of spider's web drawn
across the disk.   The breadth of some may be as
great as one or two hundred miles; of others, not
more than twenty miles.

The conclusion that Schiaparelli drew from his
long-continued observations was that these channels
were fixed configurations on the planet.   Their

length and arrangement were constant or varied only within narrow limits, and each one began and ended between the same regions. But their appearance and degree of visibility varied greatly from one opposition to another and even from one week to another. The variations in the appearance of the different channels were not simultaneous, but appeared to occur capriciously, so that one might become indistinct or even invisible whilst another in its vicinity became at the same time conspicuous. The channels intersected one another at all possible angles, but usually at the small dark spots, which Schiaparelli interpreted as lakes. Every channel opened at its end into either a sea, or a lake, or into another channel. None of them was cut off in the middle of a continent, remaining without beginning or end.

His considered conclusion in 1893 was that the *canali* were truly great furrows or depressions in the surface of the planet, destined for the passage of water. The changing appearance of the channels he attributed to inundations resulting from the melting of the snows, followed by the soaking away of the water and its eventual drying up. He added that the network of the channels was probably a geological formation and that it was not necessary to suppose them to be the work of intelligent beings.

The most surprising feature about the canals (we shall henceforth use the term that has come into general use) was their doubling. This occurred, according to Schiaparelli, who first announced it in 1882, principally in the months preceding and following the melting of the polar cap. In the course of a few days, or even of a few hours, a canal would change its appearance and be transformed throughout its length into two lines or uniform stripes, more or less parallel to one another, which ran straight and equal with the exact geometrical precision of the two lines of a railroad. The two canals, he asserted, followed very nearly the direction of the original canal and ended where it ended. One of the two might follow the course of the original canal or it might be that they would lie on either side of it. The distance between the two canals might range from about thirty miles to three or four hundred miles.

The doubling did not occur at the same time for all those canals that showed the phenomenon, but was produced here and there, in an isolated irregular manner and without any recognisable order. At different oppositions, the doubling of the same canal produced different appearances, as to width, intensity and arrangement. It was therefore concluded that the doubling could not be due to a fixed formation on the surface of Mars, of a geographical character like the canals.

Schiaparelli added: " Their singular aspect, and their being drawn with absolute geometrical precision, as if they were the work of rule or compass, has led some to see in them the work of intelligent beings, inhabitants of the planet. I am very careful not to combat this supposition, which includes nothing impossible." He went on to consider various suggested explanations that had been put forward to account for the appearance and concluded: " The examination of these ingenious suppositions leads us to conclude that none of them seem to correspond entirely with the observed facts, either in whole or in part. Some of these hypotheses would not have been proposed, had their authors been able to examine the doubling with their own eyes. Since some of these may ask me directly,—Can you suggest anything better? I must reply candidly, No."

The conclusions reached by Schiaparelli from his long-extended observations of Mars have been given in some detail because they had the effect of giving an immense stimulus to the study of Mars. The news that changes could be seen occurring on the surface of Mars was interpreted by some as providing evidence that there were intelligent beings on Mars for, it was argued, the numerous canals following straight or regular paths could not be natural formations, but must have been artificially made. This interpretation was violently combated by others; Schiaparelli, as we have seen, kept an open mind on this question; he did not accept it as proved, though he did not regard it as impossible.

The great protagonist of the theory of the artificial nature of the canals was the American, Percival Lowell, who in 1894 founded an observatory at Flagstaff, in Arizona, for the special purpose of studying the planets, and Mars in particular. The site for the observatory, at a high altitude in the dry region of Arizona, was selected for the excellence of the atmospheric conditions. There, for many years, Lowell and his assistants studied Mars assiduously, whenever it was sufficiently well placed for observation, and accumulated a mass of information about the changes that occur on its surface.

One of the early discoveries made at the Lowell Observatory was that the dark areas of the planet, which had hitherto been regarded as seas, showed a considerable amount of detail and that they were crossed, like the ruddy areas, which had hitherto been regarded as continents, by canals. Changes both of colour and form occurred in these dark areas and it was concluded that such changes were primarily of a seasonal character. These results provided conclusive evidence that the dark areas could not be seas, in the ordinary sense of the word. If they were watery areas, they must be more in the nature of marshland. The conclusion to which Lowell came was that the dark areas were regions where there was vegetation, the fertile regions of Mars, in contrast to the ruddy-coloured regions, which represented the arid desert regions, where no vegetation could grow.

The changes that were found to occur in the dark areas were of two kinds: some were quite irregular whilst others were considered to be seasonal in character. There can be no doubt that at times well-marked changes occur over considerable areas, which are not of a seasonal nature. Thus, for instance, a certain marking observed by Schiaparelli, and named by him *Lacus Mæris*, could not be found by Pickering at Arequipa in 1892. It reappeared with perfect distinctness, and was observed by Lowell, in 1903, after an interval of thirteen years since it had last been seen. Another marking, called *Lacus Solis*, showed a marked change between

1924 and 1926, being extended much farther to the north in the later year; by 1928, it had recovered its normal appearance. Such large-scale changes are relatively infrequent.

The visibility of many other markings was found by Lowell to change in a regular seasonal manner. The general character of these changes were the same. Soon after the beginning of the melting of the ice-cap in the summer hemisphere, the canals begin to become visible in the polar regions adjacent to that cap. The darkening of the canals spreads gradually and progressively towards the equator and beyond it, into the opposite hemisphere, at the rate of about fifty miles a day. At the same time occur changes of colour of the dusky markings, from a light to a darker green and later to brown and yellow. These changes were attributed by Lowell to a transference of water from the melting pole cap towards and beyond the equator, the transference being accompanied by the growth of vegetation.

These simple and regular changes, proceeding in harmony with the progression of the seasons on Mars, have not been altogether confirmed by other observers. It might be thought that regular changes, recurring with the seasons, would be easy to establish. It must be remembered, however, that at any single opposition Mars is sufficiently well placed for observation for only a few months; as the length of the Martian year is 687 days, the advance in the seasons during the time over which observations can extend at any one opposition is not very great. In order to observe Mars throughout one complete cycle of the seasons, it would be necessary to continue observations for fifteen years. Moreover, because the rotation of Mars is about 37 minutes slower than the rotation of the Earth, there is a slow falling behind in the longitudes of Mars presented centrally to the Earth at the same hour on successive nights. In consequence, any given marking is only well placed for observation for about a fortnight consecutively, during which time there may be only one or two nights with good conditions for observation. After this time the marking becomes unfavourably placed at the hours suitable for observation and cannot be observed again for about a month.

It will be clear that it is by no means easy to follow the changes in the surface markings through a complete cycle of the Martian seasons. A further complicating factor is that it is not possible to take advantage of photography to give a permanent record of the appearance of the planet at a given instant because, as we have already explained, most of the detail on the surface is too fine to be recorded by the photographic plate; the record of the appearance of the planet must, therefore, be provided by the observer drawing what he sees. A good observer is not necessarily a good draughts-man, however, and conversely the good draughts-man may not be a good observer. But however good the draughtsman, it is by no means an easy matter to portray faithfully the detail seen on the face of the planet, in its varied light and shade, with proper contrasts of colour and with the positions and sizes of the various markings shown in their proper relationship one to another. Much of the detail that the observer is striving to portray is at the limit of vision and only to be glimpsed momentarily at rare intervals. In the circumstances, it would not be surprising to find that the observer had tended at times to concentrate his attention on certain details and at other times to concentrate it on different details. The difficulty of excluding the possibility that seasonal changes in fine detail do not arise in this way, and that they are not of subjective origin, will be appreciated.

Lowell claimed also to have observed the doubling or pairing of certain canals, previously announced by Schiaparelli, as we have already mentioned. He asserted that the majority of the canals were persistently and perpetually single, but that a proportion of them at times appeared mysteriously paired, the second canal being an exact replica of the first, running by its side throughout its whole length and keeping equidistant from it, like the two lines of a railway track. The distance between the two canals of a pair varied, according to Lowell, from about seventy-five miles to 400 miles.

We shall now consider briefly the interpretation placed by Lowell upon his observations. He suggested that we must try to see ourselves as others see us. Suppose that we could do away with the clouds that at any moment cover much of the surface of the Earth and that we looked at the Earth from Mars or Venus. From such a distance the local merges into the general aspect and, at intervals of six months, an interesting and beautiful transformation would be seen to spread over the face of

the Earth—the vernal flush of the Earth's awakening
from its winter sleep. This would be revealed
through the growth of vegetation starting in temp-
erate latitudes and moving progressively toward the
pole, the tint deepening the while. The wave of
awakening on the Earth would travel from equator
to pole, whereas on Mars it journeys from pole to
equator. The difference, according to Lowell, was
of fundamental significance for the interpretation of
the changes observed on Mars. The growth of
vegetation is dependent, provided certain other
essential requirements are satisfied, on the Sun's
rays and the presence of moisture. On the Earth
water is almost always available, except in desert
regions, but, unless it is called by the Sun, vegetation
never wakes. After the Sun has departed south in
the autumn, the vegetation in northern latitudes
must await its return in the spring.

Mars, according to Lowell, was otherwise circum-
stanced. Not merely was the warmth from the Sun
needed to awaken the vegetation to growth but it
was also needed indirectly to provide the water
supply. There is no surface water on Mars and the
annual unlocking of a water supply, provided by
the melting of the snows of the polar cap, was needed
before the vegetation could commence to grow.
The growth must, therefore, start at the pole,
where the water supply first becomes available, and
it then follows the frugal flood towards the equatorial regions.

The next step in the argument was that though
vegetable life could reveal itself directly, animal life
could not. "Not by its body, but by its mind

would it be known. Across the gulf of space it could be recognised only by the imprint it had made on the face of Mars." This imprint he recognised in the canals, the long straight markings, which were just such markings as intelligence might have made. The unnatural regularity of these markings, as Lowell drew them, made it impossible to believe that they could be natural features. There was the additional fact that many of the canals converged to dusky patches at their junctions, the so-called *oases*, and passed from one oasis to another in an absolutely unswerving direction. " The observer apparently stands confronted with the workings of an intelligence akin to and therefore appealing to his own. What he is gazing on typifies not the outcome of natural forces of an elemental kind but the artificial product of a mind directing it to a purposed and definite end."

He therefore concluded that the canals were artificial channels, made by intelligent beings, to carry the melting water from the poles across the surface of the planet, passing from point to point by the shortest possible paths. It follows, if this interpretation is correct, that Mars must be a world devoid of mountains. As the water travels along these channels, the irrigation makes it possible for vegetation to spring up along their banks and where the canals meet, at the oases, are the fertile regions where the Martian beings live.

Some explanation is required of how the water manages to flow from the pole to the equator and beyond. It cannot be flowing down hill all the way; if it were down hill from the pole to the equator, the water would be running uphill after crossing the equator. Therefore, the water must have been artificially conducted over the face of the planet. There must be a great pumping system, on a scale far surpassing any of the works of man, and this in itself presupposes an advanced type of intelligence. Lowell calculated that the power required would be four thousand times the power of the Niagara Falls.

And what is the motive for this gigantic irrigation system? That is readily found. It is provided by the instinct for self-preservation on the part of a world that is becoming increasingly arid. In the struggle for existence, water must be got. The Martians saw in the growing scarcity of water the premonition of their doom. All other questions became to them of secondary importance in comparison with the vital urge to obtain water. The only places where water is in storage and whence it may be got are the poles: hence the whole economy of life on the planet must centre round making this water available for the needs of life. As the great occupation of the Martians must accordingly be that of getting water, is it to be wondered at that it is the fruits of this occupation that have revealed their existence to the eyes of man?

Only with an intelligent population, but not otherwise, would the inevitable progressive desiccation of the planet be foreseen. The water supply would not fail in a moment; it would be a slow gradual process. Local needs would urge the reaching out to a distant supply, as is already being done on our Earth for the provision of adequate water supplies for large towns and cities. The steps to greater and still greater distances would follow one by one until eventually the surface of the planet was covered with a vast network of channels, making water available and enabling vegetation to grow.

Lowell concluded his book, entitled *Mars as the Abode of Life*, with these words: " A sadder interest attaches to such existence: that it is, cosmically speaking, soon to pass away. To our eventual descendants life on Mars will no longer be something to scan and interpret. It will have lapsed beyond the hope of study or recall. Thus to us it takes on an added glamour from the fact that it has not long to last. For the process that brought it to its present pass must go on to the bitter end, until the last spark of Martian life goes out. The drying-up of the planet is certain to proceed until its surface can support no life at all. Slowly but surely time will snuff it out. When the last ember is thus extinguished, the planet will roll a dead world through space, its evolutionary career forever ended."

Such, in brief, was Lowell's theory: attractive, ingenious and logical, provided that the observational basis can be accepted. But that is where the rub came, for though there were observers of Mars, many of them possessed of moderate-sized instruments, who confirmed Lowell's observations, there were other observers who were unable to see the essential phenomena upon which his theory was based; some of these were observers of great acuity of vision and of considerable reputation, with large instruments placed where the conditions for observation were particularly favourable. The whole question of the nature of the detail to be seen on Mars and of its interpretation became a matter of violent controversy. Time has passed; the controversy has died down; there is now a general consensus of opinion about what can be seen on Mars. Let us take stock of the situation and see what we can fairly accept as established beyond the possibility of doubt.

The main argument in favour of the Martian civilisation was the artificial appearance of the network of canals shown on the maps of Mars, compiled by Lowell and others. There is no denying that the maps do have a very artificial appearance. But they are compiled from many separate drawings and all the canals shown on a map are never seen at once; only a few are visible on any one night. Nevertheless a few canals are sufficient to give a very artificial appearance. It must be emphasised, on the other hand, that though the drawings may have been most carefully made, the planet itself—if we could approach sufficiently near it—would doubtless look entirely different in its main features from any of the drawings and would not appear at all artificial.

The question at issue is not whether the canals exist or not. There can be no question about the existence of at least the most conspicuous of them. Some can be seen in telescopes of moderate size and a few have been photographed. The photographs reproduced in Plate 14, perhaps the best photographs of Mars that have been obtained, show some of the canals clearly. The question about which there has been so much controversy is whether these features of the Martian surface are the straight, narrow, sharply defined lines, which they appear to be in Lowell's drawings. Dr. Barnard, an experienced observer with a very keen eye, who had the advantage of observing with some of the largest telescopes in America, saw them as ill-defined, irregular, diffuse shadings, which were not of uniform breadth and were not always even continuous. Observing with the great sixty-inch reflecting telescope at Mount Wilson, a telescope much more powerful than Lowell had at his disposal, he said that Mars gave " the impression of a globe whose entire surface had been tinted a slight pink colour, on which the dark details had been painted with a greyish coloured paint, supplied with a very poor brush, producing a shredded or streaky and crispy effect in the darker regions." He added that " no one could accurately delineate the remarkable complexity of detail of the features which were visible in moments of the greatest steadiness." M. Antoniadi, who has studied the planets for many years with the great refractor at Meudon, agrees with Barnard. The general consensus of observational opinion, in fact, is in agreement with Barnard that the canals do not form a sharply defined geometrical network but that they are broad, diffuse and irregular in outline.

The appearance of the canals has been described by Dr. Waterfield, who is Director of the Mars Section of the British Astronomical Association, and who has observed Mars for many years. He states that " when he first started observing Mars he found that he was inclined to see more or less linear detail upon the disk. But as time passed that tendency grew less and less, and he began to see the detail more like what he now considers to be its actual appearance. It undoubtedly takes many years for the eye to become trained for the appreciation of the finest telescopic detail. Before that is accomplished, it is liable to see hazy discontinuous streaks as narrow and continuous lines, to interpret a complex system of light and dark shadings as a more or less linear geometrical pattern, and even to join up the longer and quite obvious markings by lines that do not possess any objective reality. He finds that this tendency is still liable to recur when the planet is far from the Earth and difficult to observe, when our atmospheric conditions are poor, and when telescopes of smaller size are used. Under normally good conditions he sees the canals as wide and diffuse streaks which often form the border of a more extensive shading and are generally irregular in width and sometimes discontinuous. Finally, under the very best atmospheric conditions—which occur only momentarily on comparatively few nights in the year—some of these streaks remain while others melt away into a background of more complex structure."

There is a subjective tendency for the eye to connect up detail in the form of irregular shadings and markings, which are almost at the limit of vision and glimpsed only with difficulty, by continuous lines. The following experiment is easily made. Draw a row of dots, one-eighth of an inch apart, on a piece of paper and look at it from a distance of about thirty feet. The separate dots will not be seen; they will appear to form a continuous, uniform line. Because the canals appear to be continuous or because some of them have been recorded on photographs it does not necessarily follow that they are continuous.

It must not be thought that there is any question of dishonesty or bad faith on the part of the observers to account for such extremes as the delineations of Lowell and Barnard. These two observers studied Mars for many years under favourable conditions and both were trained and experienced observers; each of them has honestly recorded the appearance of the planet as he saw it. The only possible explanation of the differences is that the observation of these faint elusive details is subject to complex personal differences. The observer looks in the telescope and an image is formed on the retina of the eye. He has to interpret what he sees and to transfer the impression in his mind into a drawing. Differences of visual acuity must certainly be an important factor. One observer may look at the image of a twin star, whose two components are separated by a very small amount, and see clearly that the star is double; another observer will see the star as a single star. Subconscious interpretation of what is faintly glimpsed may be very different for two different persons. The eye of one may tend to bridge the gap between faint details and to draw a marking as a uniform, straight, continuous line unless he can clearly see that there are irregularities, bends and discontinuities in it. Another may only draw it in this way

13

when he can see beyond the possibility of doubt that it is uniform, straight and continuous.

The truth may lie somewhere between the two extremes, though it appears probable that Barnard's delineation is the nearer approach to the truth. If the canals are as sharp and straight as Lowell depicted them, it seems incredible that other observers, if they could see them at all, should not also see them as sharp and straight. The experiment was tried of placing a diagram, on which there were scattered a number of dots of various sizes, short lines and shady patches in an irregular manner, before a class of school children, who were told to draw what they could see. Many of them, particularly those at the back of the room, connected up the more prominent features with straight lines. It is much more likely that the eye will tend to connect up disconnected structure in this way and to represent it as continuous than that it will tend to break up continuous structure into disconnected patches.

Lowell's representation of paired canals seems to provide positive proof that his observations were liable to subjective error. If two lines are drawn in ink on white paper and viewed from a distance with a 6-inch telescope, it is not possible for the eye to separate them unless the angular separation is at least one second of arc. This arises from the fact that, for optical reasons, the telescopic image of a sharp line is not perfectly sharp; the image is somewhat broadened: if there are two lines sufficiently close together, the images will converge into one another so that no eye, however acute, can possibly separate them. The limiting separation of one second of arc, for a 6-inch telescope, is inferred from optical theory and confirmed by observation. Lowell, however, with a six-inch telescope recorded paired canals whose separation was as small as 0·26 second. His own assistant expressed grave doubts about the objective reality of the duplication, and other experienced observers have not been able to confirm the existence of these paired canals.

The observations and deductions of Lowell have been described in some detail because of their great interest for the present purpose of considering the evidence for or against life on the planets. The conclusion which it seems reasonable to accept is that the geometrical network of narrow straight canals does not exist. There is no doubt that there are faint markings in the form of hazy streaks, which are fairly straight and seem to be pretty continuous. Under conditions of exceptional steadiness, they appear to resolve into finer and more complicated detail. It is doubtful, therefore, whether these markings are really continuous and it is idle to speculate about their nature. We must abandon Lowell's theory of an artificially constructed network of water channels and accept the canal markings as natural formations. The difference in the topography of the Earth and the Moon serves as a reminder that other worlds need not necessarily bear any close resemblance to the Earth in surface features. It was suggested by Dr. W. H. Pickering that the canal formations might be due to volcanic cracks lying between craterlets on the Martian surface; water-vapour might escape from these craterlets and cracks and nourish vegetation growing along their sides; it is this vegetation and not the crack itself that would be visible in the telescope.

We must lay aside all these speculations and preconceived notions, however captivating to the imagination they may be, and consider without bias what information about Mars has been substantiated by further observation and what conclusions can legitimately be drawn from it.

We consider first what direct observational evidence there is of an atmosphere on Mars. In our survey of the solar system we have so far encountered some worlds, such as the Moon and Mercury, whose surface we can see but which are devoid of an atmosphere, and other worlds, such as Venus and the major planets, where there is ample evidence of an atmosphere but where we are unable to see the solid surface. When we come to Mars, we are able to see the surface and we expect, as we have already mentioned, to find an atmosphere. The seasonal change in the size of the polar caps provides indirect evidence that there must be an atmosphere. When the polar cap melts with the advance of the summer season, there must also be some evaporation of moisture; if Mars had entirely lost its atmosphere, it must also have lost this water-vapour in the course of ages and the material of which the polar caps are formed would gradually have been completely dissipated away into space.

But we do not have to rely, fortunately, merely on indirect evidence of an atmosphere on Mars.

Direct evidence of an atmosphere was provided by Dr. Wright who, at the Lick Observatory, by using appropriate colour filters, photographed Mars by the light of different colours. Photographs by infra-red light, which has penetrating or haze-cutting properties, showed the surface detail clearly; the photographs in the ultra-violet light showed practically no trace at all of surface features. This is illustrated in Plate 12, which shows a terrestrial landscape and Mars photographed in infra-red and ultra-violet light respectively. The terrestrial landscape represents the view from the top of Mount Hamilton, looking towards the distant mountains across the intervening valley. In the infra-red photograph the detail is clearly shown, but in the ultra-violet photograph all that can be seen is a faint outline of the crest of the distant mountains. There was sufficient haze in the atmosphere to scatter the light of short wave-length to such an extent that all detail was lost. The comparison of the two photographs of Mars, which were taken on plates identical with those used for the terrestrial photographs, shows that the ultra-violet light is scattered by a Martian atmosphere to such an extent that it is not able to penetrate to the surface of the planet and out again, whereas the infra-red light, which, because of its longer wave-length, is not scattered so much, can get through the atmosphere to the surface of the planet and out again.

By employing special filters, Dr. Wright also photographed Mars with red, yellow, green and blue light and found that the surface detail became less and less distinct as the wave-length of the light used became shorter and shorter. The whole series of photographs provides a progressive sequence from the infra-red photograph in which the surface detail is clearly seen, the detail becoming progressively less distinct until in the ultra-violet photograph there is practically no evidence of any surface features.

The photographs revealed another interesting fact. The images of Mars on the photographs taken with the ultra-violet light are larger than those taken with the infra-red light. This is clearly shown in Plate 14. In the latter case we obtain an image of the solid planet itself; in the former case we have an image of the atmospheric shell surrounding Mars. The difference in size of the two images corresponds to a difference of fifty or sixty miles in the true radii, so that the atmosphere of Mars has a considerable depth. Comparison with terrestrial photographs taken under favourable conditions suggests that the Martian atmosphere, though evidently of considerable depth, is nevertheless very tenuous and that the total atmospheric pressure on Mars does not amount to more than a few per cent. of that at the surface of the Earth. The force of gravitation at the surface of Mars is only two-fifths of the force at the surface of the Earth, so that if Mars and the Earth had atmospheres which, under standard conditions of pressure, were of the same thickness, the atmosphere of Mars would extend to a much greater height than the atmosphere of the Earth.

Further confirmation of an atmosphere on Mars is provided by the occurrence of clouds, which can not only be seen in the telescope but can also be photographed. The clouds are of two types, those that appear white to the eye and those that appear yellowish. The white clouds are best seen on the ultra-violet photographs and are barely visible, if seen at all, in the infra-red photographs. Such clouds must occur fairly high up in the atmosphere because, if they were at a low level, the ultra-violet light would not penetrate to them; they must be sufficiently thin to allow the infra-red light to pass through them, otherwise they would appear more conspicuously on the infra-red photographs. These white clouds are somewhat rare; they have a tendency to begin to form at about Martian noon and to increase in size during the afternoon, as the temperature falls. The formation and growth of one of these clouds is illustrated in Plate 13. It is probable that they are produced by the condensation of moisture, as a result of the fall in temperature. They tend to be most conspicuous, therefore, at sunset, when they are near the edge of the disk. They may occasionally be seen, by direct observation of Mars in the telescope, projecting beyond the edge of the disk. It then becomes possible to make an estimate of the height of the cloud above the surface of the planet; heights up to twelve miles have been found in this way.

The second type of cloud, the yellow cloud, is more frequently seen. Such clouds are shown on the infra-red but not on the ultra-violet photographs. They must, therefore, be at a fairly low level in the atmosphere. An example of a yellow cloud is shown in Plate 13. These clouds appear

yellowish to the eye, but the lack of contrast between
them and the ruddy surface of the planet makes
them rather difficult to see. A very large area of
the planet is sometimes covered by the yellow
clouds, which obliterate partially or entirely the
underlying surface details; they may persist in the
same region for some time, occasionally for as long
as several weeks. It has been suggested that the
yellow clouds are clouds of dust raised by winds
blowing over the extensive desert areas of the
planets.

The polar caps show a surprising difference in
appearance in the photographs taken with light of
long- and short wave-length. The usual explanation of the caps is that they are surface deposits of
snow or hoar frost in the polar regions, analogous
to the snow- and ice-caps of the polar regions of the
Earth. The only other white solid substance of
which they could consist is solid carbon dioxide
(commercially known as " dry ice "). Solid
carbon dioxide volatilises at low pressures, such as
must exist on Mars, at temperatures appreciably
lower than the observed temperature of the caps
and the possibility that the caps may consist of solid
carbon dioxide can, therefore, be excluded. If the
polar caps are merely surface deposits, we should
expect them to be clearly shown on the photographs
in light of long wave-length but to be invisible on
the photographs in light of short wave-length,
whereas, contrary to expectation, they are seen
most clearly on the latter photographs. The polar
caps must, therefore, be largely, though not entirely,
an atmospheric phenomenon. It is probable that

over the polar regions there are clouds, like high cirrus clouds on the Earth, of no very great thickness, so that light of long wave-length can get through them and that there is in addition a surface deposit of snow or ice.

This deposit cannot be of any great thickness because, unlike the ice-caps of the Earth, it disappears almost entirely in the course of the summer months. It is quite easy to prove by calculation that the caps cannot be very thick. The intensity of the Sun's radiation at the distance of Mars is known and we can calculate how thick the cap must be in order that the whole amount of heat received by the cap during the time that the Sun is above the horizon, and assuming that none is lost by reflection or radiation, will just suffice to melt it. The thickness found on these assumptions is the maximum possible and is only about six feet. But most of the heat falling on the cap will be reflected or radiated back and will not be available for melting the ice; the true thickness must consequently be very much less than six feet and probably does not amount to more than a few inches on the average, except in the very near neighbourhood of the pole. The whole quantity of water that would be obtained by the melting of one of the polar caps would not be more than sufficient to fill a large lake about the size of Wales. A lake of such a size appears to be entirely inadequate to supply the large quantity of water that, according to Lowell's theory, was pumped for thousands of miles across the surface of Mars for purposes of irrigation.

And what about the composition of the Martian

atmosphere ?   Water-vapour there must undoubtedly be. Although there are no open seas on Mars,
the existence of the polar caps and their melting as
the summer advances, together with the evidence
of clouds, afford sufficient proof that the atmosphere
must contain water-vapour.   The amount of the
water-vapour is so small, however, that it can be
detected only with the greatest difficulty.   The
attempts to detect it have almost invariably ended
in failure.   At the Lowell Observatory, which is in
Arizona at a height of 7,250 feet above sea-level,
Dr. Slipher, in 1908, by comparing the spectra of
Mars and the Moon, when at the same altitude
and under conditions of exceptional atmospheric
dryness in the winter, found that the water-vapour
absorptions were slightly stronger in the spectrum
of Mars than in that of the Moon.   This slight
difference in intensity must have been produced by
the water-vapour in the atmosphere of Mars.
Usually the absorption by water-vapour in the
Earth's atmosphere is so strong that the much more
feeble absorption in the atmosphere of Mars is
entirely masked.

All attempts to detect oxygen in the atmosphere
of Mars have been unsuccessful, and it can be concluded that the amount of oxygen is not more than
one-thousandth part of the amount in the Earth's
atmosphere.   Indirect evidence of oxygen is provided by the ruddy colour of Mars, which is unique
amongst the heavenly bodies.   This red colour is
suggestive of rocks that have been completely
oxidised and it may be contrasted with the grey or
brownish colour of the rocks on the Moon, which

have remained unoxidised because of the absence of oxygen. It appears probable that Mars may be a planet where the weathering of the rocks, followed by their oxidation, has resulted in the almost complete depletion of oxygen from the atmosphere.

Carbon dioxide has not been detected in the Martian atmosphere. This is not surprising because carbon dioxide must be present in large quantity before it will produce absorptions of sufficient strength to be detected.

The temperature of Mars is much in accordance with what we should expect to find for a planet somewhat more remote than the Earth from the Sun. In the Martian tropics the temperature rises well above the freezing-point at noon and may reach 50° F. or a little more. The dark areas are somewhat warmer than the reddish areas. The observed temperature of the polar caps in winter is very low, about − 70° C., corresponding to about 125 degrees of frost on the Fahrenheit scale. This temperature possibly refers to the upper surface of the cloud layer above the pole, in which case the temperature at the surface may be appreciably higher. At midsummer, the temperature at the poles rises somewhat above freezing-point.

In the afternoon, as the Sun gets lower, the temperature falls very rapidly. This is because there is only a scanty atmosphere, with very little water-vapour, to act as a blanket, and prevent the escape of the long-wave heat radiation from the surface rocks, which have been heated by the Sun during the day. Water-vapour is very effective in preventing the escape of the heat radiation from the surface of a planet.   Anyone who has lived in a moist tropical climate knows that there is very little fall of temperature at night, whereas in dry desert regions, though the day temperature may be much higher than in moist tropical regions, the night temperature is much lower because of the rapid fall of temperature after sunset.   The maximum temperature on the Earth does not usually occur at noon, when the Sun is highest in the sky, but during the afternoon.   This is because of the blanketing action of the water-vapour in the atmosphere, which prevents the rapid escape of radiation from the heated surface of the Earth and causes the temperature to continue to rise for a few hours after the Sun has reached its greatest altitude.   But on Mars the temperature is highest at noon and begins to fall immediately afterwards.   By sunset the cold has become intense; the minimum temperature at night is about $-130°$ F.   The diurnal range of temperature between noon maximum and night minimum is thus very great, being about equal to the difference between the freezing- and the boiling-points of water.

The climate of Mars may be said to resemble that of a clear day on a very high mountain.   The solar radiation by day is rarely tempered by cloud or mist.   By night there is rapid radiation from the soil into space and intense cold.   The climate is one of extremes.   The changes of temperature from day to night and from one season to another are very great.   The seasons are longer than on the Earth and their length accentuates the difference between summer and winter conditions.   The seasonal changes are more pronounced in the southern hemisphere than in the northern. The distance of Mars from the Sun varies by as much as twenty-six million miles in the course of its orbital passage round the Sun. Mars is nearest to the Sun when it is winter in the northern hemisphere and summer in the southern hemisphere; it is at its farthest when it is summer in the northern and winter in the southern hemisphere. The southern hemisphere, therefore, has a warmer summer but a colder winter than the northern.

We have been compelled to discard the evidence on which Lowell based his theory that a race of intelligent beings exists on Mars. May there not be sufficient evidence, however, to enable us to conclude that there is life of some sort, but not necessarily intelligent life, on Mars? The temperature conditions are neither so high nor so low that the possibility of life can be excluded, though the great daily range of temperature and the rapidity of the changes would prove very trying for any form of life with which we are familiar. Water-vapour is certainly present in the atmosphere and there is evidence of oxygen, though the supply may be approaching exhaustion. There seems to be no reason why life on Mars could not have adapted itself to such conditions.

That there are changes from time to time in the Martian surface we have already mentioned. Some of these changes appear to be seasonal; others are quite irregular. Lowell claimed to have established a wave of darkening spreading equatorwards as the ice-cap of the summer hemisphere melted. These claims have not been altogether confirmed by other investigators, who find the changes neither so simple nor so clear cut. There is general agreement, however, that there are complete changes both in appearance and colour of various markings, which correlate with the seasonal changes. It is difficult to interpret these changes in any other way than by the seasonal growth of vegetation. The vegetation covers the dark regions of the planet, the rest of the surface being desert. As the ice-cap melts, the moisture reaches lower latitudes, possibly in the form of rivers or streams, but more probably as rain or dew. With the coming of the moisture, the vegetation commences to grow and the colour of the areas covered with plant growth changes to green. When winter comes on, the green colour gradually gives place to grey or brown.

It has been suggested by Arrhenius that the soil in the dark areas is saturated with soluble salts, like the alkali flats or salt pans that are found in some of the desert areas of the Earth. These salts are hygroscopic and absorb moisture from the air when the melting of the pole-cap is in progress; the darkish mud formed when the salts deliquesce would account for the change in the appearance of the dark areas. He suggests that in this way the changing tints could perhaps be accounted for in the absence of vegetation. The suggestion does not seem to provide a satisfactory explanation of the sequence of colour changes recorded by experienced observers of Mars.

The irregular changes in appearance can be

attributed to modifications from year to year in the growth of vegetation, possibly produced by localised climatic variations from year to year. A particular region of the planet may receive one year a supply of moisture that is fully adequate for the scanty needs of the Martian vegetation; but another year, the supply over the same region may be insufficient for these needs and, because of the drought, there will be a more or less complete failure in the growth of vegetation. It is not to be expected that on Mars, any more than on the Earth, the climatic conditions at any particular place will be exactly the same each year.

As we have mentioned, the colour of the surface of Mars provides sure evidence of the presence of free oxygen, at any rate in the past. The presence of free oxygen almost certainly demands the existence of vegetation. Combining this argument with the evidence from the changes that occur on the surface, we may conclude that it is almost certain that there is some form of vegetation on Mars.

We cannot say whether animal life, and in particular whether higher forms of life, can exist on Mars. The small amount of oxygen on Mars seems to make this improbable, though we know so little about life that the possibility cannot be entirely excluded. But the question whether or not such forms of life exist on Mars at the present time is surely of minor significance in comparison with the very strong evidence that life of some kind is to be found there. We started out on our survey of the bodies of the solar system not knowing what we

PLATE 14

THE PLANET MARS

The upper portion of plate shows photographs of Mars taken on 1939, July 20 (upper left); July 23 (upper right); August 11 (lower left); August 31 (lower right). These photographs were obtained by Dr. Jeffers with the 36-inch refractor of the Lick Observatory, through a yellow screen, so that the contrast in the photographs is similar to that seen in visual observation of Mars. The bright cap at the south (upper) pole; the dark areas (regions of vegetation); and the light areas (desert regions) are well shown. The photographs show also finer markings which some observers have interpreted as artificial canals.

The lower portion of the plate provides direct evidence that the atmosphere of Mars is of considerable depth. Images of the planet in ultra-violet and infra-red light are shown; below these the opposite halves of the two images have been juxtaposed to show that the ultra-violet image is the larger. The infra-red image, which shows the surface details, represents the solid globe of Mars; the ultra-violet image, in which no surface detail is visible, represents the solid globe surrounded by the atmospheric shell. The difference in size of the images indicates that the atmosphere of Mars has a depth of at least 50 miles.

PLATE 15

THE SPIRAL NEBULA, MESSIER 101, IN THE CON-STELLATION OF THE GREAT BEAR

The objects termed spiral nebulæ are stellar universes, lying outside our stellar universe, with which they are closely comparable in size and mass. The system, Messier 101, is seen broadside on, so that its spiral structure is clearly seen. Like our own stellar universe, it is slowly rotating in space.

If our own universe could be viewed broadside on from a great distance, it would appear generally similar to this system. The condensations of stars in the spiral arms correspond to the star clouds in the Milky Way. Patches of bright nebulosity—glowing gaseous matter—and dark patches, caused by absorption of light by fine dust, may be seen; similar features are widespread throughout the Milky Way regions of our own Universe.

The distance of this system is about $1\frac{1}{4}$ million light-years.

Photograph by Dr. Ritchey with the 60-inch reflector of the Mount Wilson Observatory, 1910, March 11. Exposure $7\frac{1}{2}$ hours.

should find but with the expectation that if conditions were suitable for life on any world, life
would somehow have come into existence there.
Our quest for conditions suitable for life to exist
was unsuccessful until we came to Mars; on one
world after another the conditions were found to be
such that we could say with reasonable certainty
that life could not possibly exist.   Finally we came
to Mars, when at length we found a world where
the conditions were such that the possibility of life
of some sort existing could not be excluded.   And
there we find clear evidence of changes taking place
which we can only attribute to the growth of
vegetation.   If this conclusion is accepted, it follows
that life does not occur as the result of a special
act of creation or because of some unique accident,
but that it is the result of the occurrence of definite
processes; given   the   suitable   conditions,   these
processes will inevitably lead to the development of
life.

But Mars, though the home of life, is a dying
world.   It has lost most of its atmosphere; it has
lost most of its moisture.   It may in the past ages
have been the home of animal, and conceivably of
intelligent, life.   It does not seem that the present
scanty supply of oxygen can be nearly adequate to
maintain such life.   Animals need oxygen to supply
energy, through the process of combustion, which
enables them to maintain the vital processes.
Evolution may secure adaptation to gradually
changing conditions but, with a progressively
lessening supply of oxygen, there must come a time
when adaptation can do no more and the moulding

powers of evolution succumb before an unequal struggle.   So it seems to me that we must look upon Mars as a planet of spent life.   Such vegetation as now continues to maintain a precarious existence must be doomed to extinction in a time which, geologically, is not remote.

In Venus we saw a world where conditions are probably not greatly different from those that existed on the Earth many millions of years ago. In Mars, on the other hand, we see a world where conditions now exist which resemble those that will probably prevail on our Earth many millions of years hence, when much of our present atmosphere will have been lost.

CHAPTER IX

THE ORIGIN OF
THE SOLAR SYSTEM

WE have now concluded our survey of the solar system. With the significant exception of Mars, the search for any evidences of life has given a negative result. We have necessarily had to rely to a considerable extent upon the inferences that can be drawn from the general conditions prevailing on the other worlds that we have considered. The only world on which we might expect to see some direct evidence of animal life, if it existed, is the Moon. The surfaces of Venus and the major planets are permanently hidden from us by clouds and it would not be possible to obtain any direct evidence even of vegetation on these worlds, if such existed. The general considerations about the nature of life have served as a guide and our conclusions should be reasonably well established. The discussion of the conditions prevailing else-where in the solar systems has provided us with information that will be useful in trying to assess the probability that life may exist elsewhere in the universe.

We have so far dealt only with the family of our Sun, which is merely an average star, one amongst the many stars, numbering a hundred thousand millions or so, in our stellar universe. And that universe in turn is merely one amongst many millions of more or less similar island universes—each a gigantic system—scattered through space.

The question that naturally suggests itself is: What is the chance that life exists on some of the planets belonging to one or other of these innumerable stars in our own universe or in some other universe? This is a difficult question to answer, because if such planetary systems exist, they are far beyond our range of vision. If the nearest known star, twenty-five million million miles away, had a planet belonging to it of the size of Jupiter—which is much the largest of the planets in our solar system—we should not be able to see it. We must give up hope, therefore, of ever being able to see any of the planets that may belong to other stars. To learn anything about the conditions prevailing on such worlds is accordingly entirely out of the question.

But we can perhaps obtain some guidance from general considerations. If we can find out how the solar system came into being we shall possibly be able to judge what likelihood there is that other stars may have families of planets. For the solar system has certainly not come into existence as the result of chance. It is not an accidental collection of bodies: there are too many regularities in the system. Let us look briefly at some of these regularities.

In the first place, all the planets, including some 1,500 minor planets, small bodies whose orbits lie between the orbits of Mars and Jupiter, revolve round the Sun in the same direction and their orbits lie nearly in the same plane. The inclinations of the planes of the orbits of the planets to the ecliptic, the plane of the orbit of the Earth, are mostly only a few degrees. With the exceptions of Pluto and of some of the asteroids, the largest inclination is 7° for the orbit of Mercury. The inclination of the orbit of Pluto is 17° and some of the minor planets have still large inclinations, but the orbits of these bodies may have suffered considerable perturbations. The orbits, again with the exception of those of Pluto and of some of the minor planets, are nearly circular in shape. The Sun rotates in the direction in which the planets revolve and the orbits of the planets lie nearly in the plane of the Sun's equator; the planets also rotate in the same direction as the Sun. The orbits of the satellites, with a few exceptions, are nearly circular and lie nearly in the plane of the parent planet's equator. The satellites revolve around the planets in the same direction as that in which the planets themselves rotate, and their movements are therefore in the same sense as those of the planets. We may compare the solar system with a pack of cards as it comes from the makers, in which the cards are arranged in suits and each suit in order of value, rather than with a pack which has been used and shuffled. The regularities demand an explanation, but to find an explanation of how a system such as the solar system could have come into being has proved to be the most difficult of all the problems of cosmogony.

The oldest hypothesis that we need mention seems first to have been suggested by that versatile Swedish scientist and theologian, Emanuel Swedenborg and, somewhat later but probably independently, by an Englishman, Thomas Wright, of Durham. It was adopted by the German philosopher, Immanuel Kant, in his essay on the general history and theory of the heavens, published in 1755. Kant started from the assumption that the material, which now forms the Sun, the planets and their satellites, was formerly a diffuse nebula. He supposed that the different attracting powers of the various elements of this nebula would cause a loss of homogeneity; the heavier elements would tend to fall to the centre but their fall would be opposed by the tendency of the gas to expand. He imagined that in this process lateral movements would be set up and that from these movements there would ensue a rotation of the whole mass. Kant was in error in supposing that rotation could be started in this way; this assumption is in direct opposition to one of the general principles of mechanics, the principle of conservation of angular momentum.

As this principle is of great importance in discussing the origin of the solar system, a few words in explanation will not be out of place. Without attempting any precise or strictly logical definition, the angular momentum of a body is a measure of the total quantity of rotational motion that it contains. Suppose, for instance, that the body is spinning around an axis, like the Earth. Imagine it to be divided up into equal parts, each weighing one pound. The angular momentum of any of these parts can be measured by the product of its velocity and its distance from the axis. The angular momentum of the whole body is the sum of the angular momentum of all the parts. Consider, now, the Earth, which rotates on its axis once in the course of a day. If the Earth were to expand, the distance of each part from the axis would be increased. As the total angular momentum must remain constant when no external forces are acting on the body, the velocity of each part must decrease to compensate for the increased distance from the axis; in other words, the rotation must be slowed down. The day would therefore become longer. Kant, in his theory, supposed that a general rotation developed out of local rotations. But these local rotations must have been some in one direction and some in another and the net effect was nil, for there was no rotation and no angular momentum to begin with. The total angular momentum must therefore have remained zero, so that the local rotations could never give rise to a general rotation.

Kant's theory was revived in a modified and more scientific form by the great French geometer, Pierre Simon de Laplace, the Newton of France, in his *Exposition du Système du Monde*, published in 1796. Laplace avoided the difficulty of explaining how the rotation had been started by postulating that the original diffuse nebula was itself in slow rotation. As the nebula cooled it gradually shrank and became more dense; at the same time, its rate of rotation necessarily increased, in order to conserve its angular momentum. With increase in the rate of rotation the centrifugal force at the equator increased until at length it became equal to the force of gravity. When this occurred, Laplace supposed that a ring of matter split off round the equator. The main mass went on contracting still further and in due course another ring would be shed and so on. The successive rings somehow condensed to form the planets; the planets themselves passed then through a similar type of evolution, contracting and throwing off rings, which in turn formed the satellites.

Laplace's exposition, though clothed in mathematical language, was semi-popular and purely qualitative. He did not discuss the problem in a mathematical or quantitative way to prove that the evolution of the primitive nebula must follow the course he suggested. The theory is attractive because it readily accounts for the motions of the planets and of their satellites being in the same general direction and nearly in the same plane, and for a long time it was accepted as providing a satisfactory explanation of the main features of the solar system. Unfortunately for the theory, however, there are objections to it that are fatal.

In the first place, Laplace assumed that a ring of matter would be thrown off around the equator of the contracting mass and that this ring would coalesce into a single body. He offered no proof of this assumption. On the other hand, it was proved by strict mathematical arguments by Clerk Maxwell in 1859 that coalescence into a single body could not occur, but that a ring of small bodies moving around the parent mass in similar orbits would result and that such a ring would form a stable configuration. The rings of Saturn provide an example of such a stable system consisting of a large aggregation of small discrete bodies.

This objection is in itself sufficient to disprove Laplace's hypothesis in the form in which it appears in the *Système du Monde*.   But there is a further and equally serious objection.   When we evaluate the angular momenta of the different members— Sun, planets and satellites—of the solar system, we find that Jupiter contributes more than half the total amount and that the four major planets between them account for about 98 per cent. of the whole.   The remaining 2 per cent. is provided almost entirely by the rotation of the Sun itself, the four inner planets—Mercury, Venus, the Earth and Mars—together contributing only 0·1 per cent. of the total.   Thus almost the whole of the angular momentum of the system is contributed by the four major planets, which have less than one-seven-hundredth of the total mass of the system. We know by the principles of mechanics that the angular momentum of the solar system at the present time must be equal to the angular momentum of the primitive nebulous mass, from which it is supposed to have been formed.   The problem is to explain how nearly the whole of the angular momentum was captured by such a small portion of the whole system.

We can look at the problem in another way. Suppose the Sun and all the planets were united to form a spherical nebulous mass of uniform density extending to the orbit of Pluto. We suppose that this mass is rotating with the period of revolution of Pluto, for a condensation within the mass must be assumed to have revolved with the same period as the mass itself. The angular momentum of such a system is found to be less than one-two-hundredth part of the actual angular momentum of the solar system. Since, in the process of contraction supposed by Laplace to have occurred, angular momentum could neither have been lost nor gained, it is certain that the present distribution of angular momentum could not have been produced in the way that Laplace supposed. The only possibility of accounting for the present distribution of angular momentum is to suppose that there was some process by which the angular momentum of the major planets was imparted to them from outside the Sun or primitive nebula by some force of a transitory nature. The origin of the solar system is not to be explained by the gradual cumulative action of internal forces; an explanation must be sought in the swift catastrophic action of forces from outside.

Attempts have accordingly been made to account for the origin of the system by supposing that at some time in its past history another star passed very close to the Sun, narrowly avoiding an actual collision. Let us consider what sequence of events would be likely to occur. As the other star approached the Sun, its gravitational attraction would distort the Sun by raising a tidal protuberance on it; the Sun would raise, in turn, a tidal protuberance on the approaching star. As the distance between the Sun and the star became smaller the tides raised on the two bodies would gradually increase more and more. Suppose the passing star did not pass nearer to the Sun than a couple of million miles or so, the tide raised on the Sun would reach its maximum height when the star was at its nearest to the Sun and would then fall back as the star moved away, causing an oscillation or pulsation in the Sun which would gradually be damped out by friction.

Suppose, however, that the star passed so close to the Sun that its attraction on the nearest portion of the Sun's surface was greater than the force of the Sun's gravity; material would then be drawn out from the top of the tidal protuberance by the gravitational attraction of the passing star. The ejection of matter from the Sun would occur slowly at first but at a gradually increasing rate as the star drew nearer. The ejection would be most rapid when the star was at its nearest to the Sun and would become progressively less rapid as the star moved away, until at length it ceased altogether. Some of the material torn from the Sun would probably be captured by the passing star and some would probably escape into space. But some would remain under the control of the Sun's gravitation and it is supposed that the planets ultimately condensed out of this material.

These assumptions have formed the basis of several theories of the origin of the solar system. Such theories have the advantage over the theory advocated by Laplace that the angular momentum of the planets can be attributed to the action of the passing star. The ejected material would form a jet curved towards the star and the angular momentum of this jet is derived from the angular momentum of the star. After the encounter, the ejected material would be moving in the plane of the path of the star past the Sun and in the same direction round the Sun. The theories of this type thus account in a perfectly natural manner for the orbits of the planets and of their satellites being nearly in one and the same plane and for the direction of their revolutions being all the same. Closer examination is needed before we can say whether such theories are free from objection when considered quantitatively; they seem to be satisfactory when we consider them merely qualitatively.

The first theory of this nature was proposed about forty years ago by Professors T. C. Chamberlin and F. R. Moulton and is known as the " planetesimal theory." Observations of the Sun show that its surface is in a state of continual disturbance and that, from time to time, incandescent material is thrown up with great force from the surface to considerable heights. Chamberlin and Moulton supposed that such disturbances, under the action of the gravitational pull of the passing star, resulted in huge eruptive bolts of matter being ejected from the Sun with great violence. From the opposite side of the Sun, where the attraction of the passing star was much smaller, another series of bolts, but of smaller size, were ejected. They supposed that the larger bolts ultimately gave rise to the major planets and the smaller bolts to the terrestrial planets.

The material that was shot out from the Sun by the eruptions, enormously intensified by the tidal forces produced by the approach of the other star, would rapidly cool. Before long it would liquefy, to form a large number of separate small bodies, each moving around the Sun in its own orbit almost independently like a planet. Soon afterwards these would solidify. The separate small bodies were called *planetesimals*. Their distribution would not be uniform; the densest portions of the bolts would give rise to close aggregations or swarms of particles. It was supposed that these collected into solid cores, which became the nuclei of the planet. As these nuclei moved through the swarms of planetesimals, they gradually gathered them in, one by one, through the action of their gravitation, until at length the neighbourhood of the Sun had been swept clean. The numerous minor planets or asteroids arose from a mass of material which was deficient in any large nucleus, so that there was nothing to sweep them up. The satellites of the planets are supposed to have been formed from smaller secondary nuclei, which were saved from being drawn in to the larger nuclei by possessing a sufficient velocity to enable them to revolve around the parent planetary nucleus, just in the same way as it was the motion of these nuclei that kept them from falling into the Sun.

Some of the material that had been drawn out from the Sun would, of course, fall back upon it, as the result of the gravitational pull of the Sun. This material had acquired angular momentum, in the same direction as the planets, from the passing star. When it fell back into the Sun, its angular momentum was communicated to the Sun, which in this way acquired a rotation in the same direction as that in which the planets revolve round the Sun. A similar explanation can be used to account for the axial rotations of the planets.

At the time that the planetesimal theory was put forward, very little was known about the propulsional forces that were responsible for the ejection of material from the Sun, seen in some of the eruptive prominences. It is now known that these forces arise from the pressure of the Sun's radiation; though this pressure can impart high velocities to particles of molecular dimensions, it is not competent to play the important rôle of a trigger releasing bolts of matter sufficiently massive to form the planets. This objection is avoided in a modified tidal-ejection theory proposed by Sir James Jeans.

Jeans showed that the effect produced by the passage of the star past the Sun would depend upon the physical nature of the Sun. If the Sun were of uniform density throughout and incompressible—in other words, very much like a solid body—the effect would be to disrupt it into pieces comparable in size and mass with the parent body. If, on the other hand, the density of the Sun increases rapidly inwards, the disrupted fragments are small, and the mass of the Sun would not have been much affected. The ejected matter would have come from the outer layers of relatively low density. The density of the Sun must increase rapidly from the surface inwards towards the centre and the conditions therefore approximated to the second case. This being so, the passing star

SPIRAL NEBULA IN THE CONSTELLATION OF BERENICE'S HAIR

The object reproduced on this plate is a spiral nebula seen almost exactly edge on. It illustrates the aptness of Herschel's description of our stellar universe as being a much-flattened system, shaped like a grindstone. The bright central nucleus will be noted, as well as the dark absorbing matter scattered through the median plane. Such absorbing matter, fine dust, is widely scattered through the central regions of the Milky Way, causing the Milky Way through a great part of its extent to appear to have two branches; the central portion is not seen through the fog of dust.

If our own universe were viewed edgewise-on from a great distance, it would appear like this nebula. The Sun has a position about midway between the central nucleus and the outer edge.

The distance of this system is about 6½ million light-years.

Photograph by Dr. Ritchey, with the 60-inch reflector of the Mount Wilson Observatory, 1910, March 6–7. Exposure 5 hours.

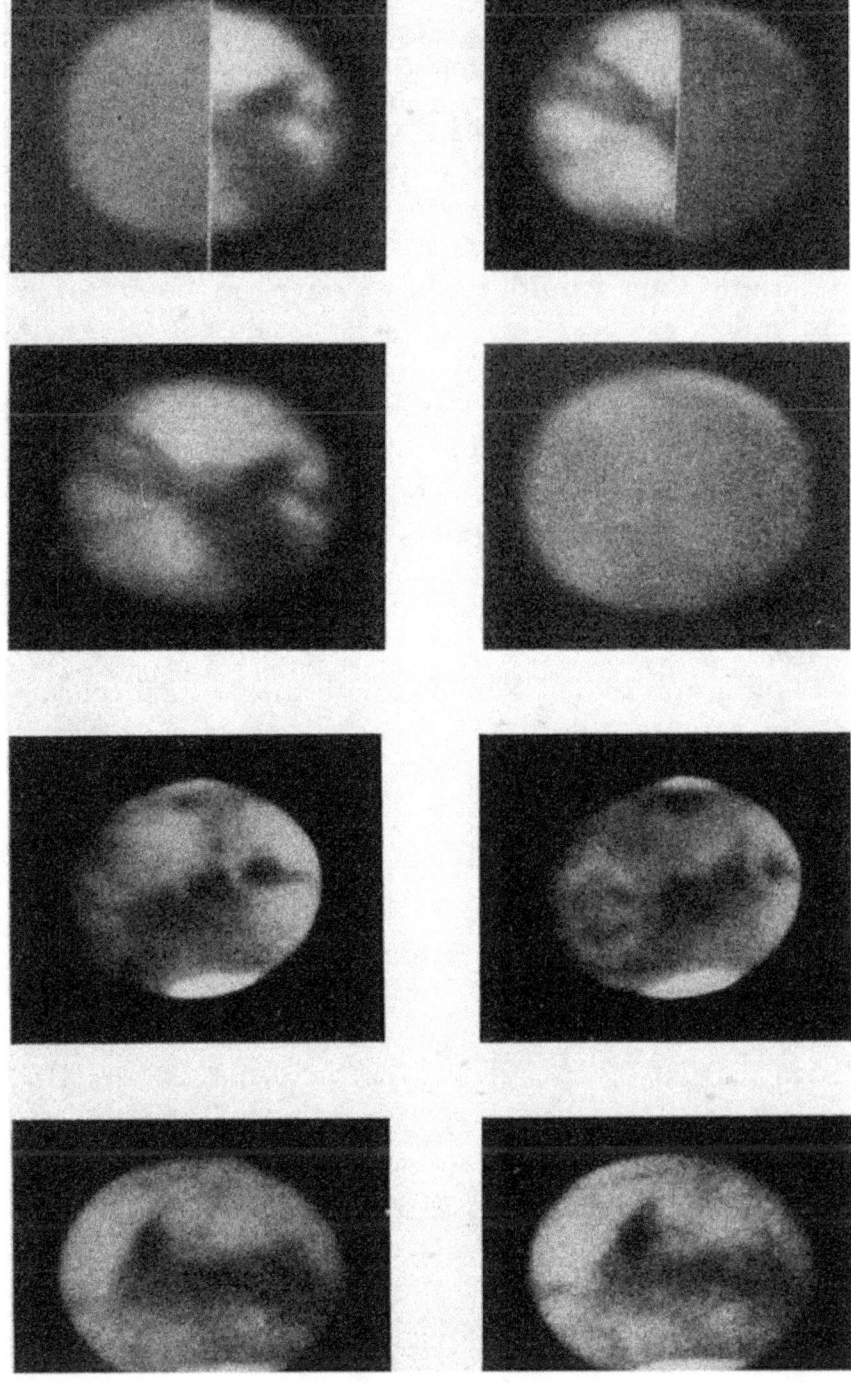

PLATE 17

SPIRAL NEBULA, MESSIER 81, IN THE CONSTEL-LATION OF THE GREAT BEAR

The spiral system is seen obliquely. Other systems can be seen at all angles of view, from broadside-on to edgewise-on, establishing the general similarity of the objects shown in the two preceding plates, though at first sight they would seem to have no points of resemblance.

The nebula, Messier 81, has a very bright nucleus. Aggregations of stars, in the form of star clouds, can be seen in the outer extensions of the spiral arms; patches of dark nebulosity, caused by obscuring dusty matter, and of bright gaseous nebulosity are clearly visible in the inner regions.

The distance of this system is about $2\frac{1}{2}$ million light-years.

Photograph by Dr. Ritchey, with the 60-inch reflector of the Mount Wilson Observatory, 1910, February 5. Exposure $4\frac{1}{4}$ hours.

would give rise to a tidal protuberance, from whose tip matter would be drawn out in the form of an elongated jet or filament. Jeans showed that unless this filament were of uniform density, which would have been extremely unlikely, there would be a tendency for the material to condense around any portions of greater density. The filament would therefore break up under its own gravitation and form a number of separate bodies which, after the disturbing star had gone its way, would move in orbits around the Sun. As the outflow of matter would be greatest when the star was at its nearest, the filament would be thickest near its middle, tapering off towards both ends. It is accordingly to be expected that the largest planets will be found in the median range of distance, with smaller planets at the two extremes of the range. This is exactly what we find in the solar system, Jupiter and Saturn being the largest and most massive planets whilst Mercury and Pluto are the smallest and least massive.

As in the planetesimal theory, there would be a portion of the ejected material which would not be drawn in by gravitational attraction to any of the planetary aggregations. This would form a mixture of gaseous matter and solid particles. On either this theory or the planetesimal theory the motion of the planet through this material, which would act as a resisting medium, would make their orbits less elliptical and more nearly circular. At the same time the planets would gradually sweep up this material. The nearly circular shape of the orbits of the planets can thus be explained.

It is supposed that the satellites of the planets were formed in an analogous manner. The passing star or the Sun, or possibly both, would have produced tides on the primitive planets, causing filaments of matter to be ejected which condensed into satellites. Just as some of the ejected material after falling back on to the Sun is supposed to have set it into rotation, so some of the material ejected from the planets is supposed to have fallen back on to their surface and to have set them into rotation.

But this raises a difficulty. To explain the rapid rotation of Jupiter it is necessary to suppose, as was shown by Dr. H. Jeffreys, that the material falling back on to its surface had a mass about one-fifteenth of the mass of Jupiter, or about 400 times the mass of all its satellites. It is not possible to believe that so small a proportion of the ejected material would condense to form satellites and that the bulk of the material would fall back to the surface of the planet. There are similar difficulties in accounting for the rotations of the other major planets—Saturn, Uranus and Neptune—and of Mars.

Dr. Jeffreys showed that this difficulty could be avoided if somewhat more specialised conditions for the approach of the star to the Sun are assumed. Instead of supposing that the star passed the Sun at a distance of a couple of million miles or so, he supposed that there was an actual collision. The picture of what happened is then somewhat as follows. As the Sun and star drew near to one another, their velocities rapidly increased, under the influence of their rapidly increasing mutual gravitation, until at the instant of collision their relative velocity attained the enormous value of several hundred miles a second. The collision was neither a direct collision nor a grazing one, but sufficiently tangential for the heavy central cores of the two bodies to miss one another.

Whilst the star was drawing rapidly near to the Sun, enormous tidal distortions would be raised both on the Sun and on the star, from which ejection of matter would begin to occur shortly before the actual collision took place. When the collision occurred, the outer layers of the two bodies within the area of impact would intermingle, whilst the heavy central portion of the star would pursue its headlong way and, swinging round in a hyperbolic path, would recede away into space.

The intermingling layer, greatly compressed by the impact between the two bodies and intensely heated by friction, would be thrown into an extremely turbulent state, with rapid rotation caused by the shearing motion of the two bodies. As the star moved away, this rapidly rotating layer would be stretched out all the way from the Sun to the star and would form a ribbon or filament, which would take the place of the tidal filament in the theory of Jeans. The whole of this stupendous catastrophe, leaving its permanent mark on the Sun in its system of planets, would be over in about an hour.

Jeffreys showed that this hypothesis of an actual collision would account satisfactorily, not merely qualitatively but also quantitatively, for the rates of rotation of the Sun and planets and for the total mass of material forming the planets.

But in removing one difficulty we have introduced others of a different nature, though equally serious. The mass of the Sun was not much affected by the collision, which merely tore off a part of its outer layer of low density, leaving the heavy central core unaffected. The luminosity and size of a dwarf star like the Sun is determined primarily by its mass and it therefore follows that at the time of the collision the Sun was practically in its present state and, in particular, its size was near about what it now is. The temperature of the highly compressed and rapidly rotating layer from which the planets are supposed to have been formed must have been of the order of ten million degrees. At so high a temperature, the average velocities of the atoms are so great that the attraction of the filament could not hold the material together. It would rapidly diffuse away into space. It is difficult to comprehend how bodies of the size of the major planets could have condensed out of such a filament.

Suppose, however, for the sake of further argument that this difficulty could be overcome. We still have to face an even more serious difficulty, whether we suppose that there was an actual collision between the Sun and the star or merely a close approach. This difficulty is concerned with the distribution of angular momentum in the solar system. If the amount of the angular momentum per ton of matter is calculated for each of the planets, we find that there is a steady increase outwards from the Sun. If we take the angular momentum per ton for the Earth as the unit in which to express the values of the angular momenta per ton of the other planets, the values increase from 0·6 for Mercury to 6·1 for Pluto. Now compare these values with the angular momentum per ton of the star. Suppose that the star was of about the same size and mass as the Sun and that at its nearest approach it was about 1,500,000 miles from the centre of the Sun. Its angular momentum per ton can be shown to be about 0·25 in the same unit. If, on the other hand, the star had actually collided with the Sun, the angular momentum per ton is found to be much smaller even than this value.

The average angular momentum per ton of the planets is about ten times greater than the average angular momentum per ton of the intruding star, if there was no collision. If a collision had occurred, the discrepancy becomes even greater. The difficulty is to explain how so much angular momentum could have been put into the matter ejected from the Sun.

The difficulty involved can perhaps be better realised in this way. The angular momentum per ton depends upon two quantities: the distance of the matter from the Sun and its velocity at right angles to the direction to the Sun. If the matter is moving directly towards or away from the Sun, it has no angular momentum about the Sun. Now the velocity of the star was such that it flew in a hyperbolic orbit away into space; if any of the ejected matter had moved in the path of the star, with the same velocity as the star, it would have done the same and would not have condensed into planets. How then has it happened that the matter that did condense into planets has acquired angular momentum (per ton) so much greater than the star and yet has not escaped into space but has remained bound by the Sun's gravitation, whilst the star was able to escape and has not remained held by the Sun's gravitation to form a double star? We must remember that the matter, which subsequently condensed into the planets, is supposed to have been drawn out initially in a curved jet, stretching from the Sun towards the passing star. To have the requisite angular momentum we need the matter to be moving at right angles to its direction to the Sun, instead of nearly in a radial direction.

A mathematical discussion of the problem is needed for the difficulty to be fully appreciated, and such a discussion would be beyond the scope of the present book. The discussion shows that it is impossible to account for the angular momentum per ton of matter of the planets being so much in excess of the angular momentum per ton of matter of the star, whatever assumption is made about the speed and direction of approach of the star. Attractive as these theories of the origin of the solar system seemed at first sight, we are reluctantly compelled to discard them as unworkable. The hypothesis of an actual collision, introduced to provide a satisfactory explanation of the rapid rotations of some of the planets, proves to raise greater difficulty than the hypothesis of a close approach, when we consider the angular momenta of the planets per ton of matter.

In order to avoid this very serious difficulty, a further modification of the theory has been attempted. Attention may here be drawn to the circumstances that each new objection raised against any theory of the origin of the solar system has to be overcome by the introduction of some new additional assumption, making the theory in itself less probable. But to account for the solar system, as it now exists, is so beset with difficulties that we cannot reject any theory, however improbable it may seem, if it gives a plausible explanation of how such a system could come into existence and if it encounters no serious objections. The solar system must have had an origin; if we cannot account for it except by the introduction of many special and somewhat artificial hypotheses, we shall have to conclude that the probability of other stars having systems of planets is very small.

In the new modifications of the theory, suggested by Prof. H. N. Russell and developed by Dr. Lyttleton, it is supposed that before the encounter with the passing star, the Sun was a twin star. This is not in itself a very improbable assumption, for it is known that a considerable proportion of the stars, perhaps one in five, is a twin system. It is suggested that the Sun's companion was a good deal smaller than the Sun and that its distance from the Sun was comparable with the distance of the major planets. The passing star, which was much more massive, is supposed to have collided with the companion to the Sun, breaking it into fragments and knocking it out of its orbit. The assumption that, if two stars collided, one of them would be broken up into several fragments of comparable size is not capable of verification by mathematical discussion and may not be correct. This is a weakness of the theory, which we must tentatively pass over.

The passing star is assumed to have made an almost central collision with the companion and to have carried away with it most of the fragments, leaving merely the debris of the collision as the material which remained associated with the Sun to form the planets. Moreover, unless the plane of the orbit of the original companion about the Sun and the path of the intruding star were nearly parallel, the orbits of the fragments that remained after the collision would not be nearly in one plane and could not therefore lead to a system like our solar system. The theory thus abounds with special assumptions. But it gets round the two difficulties of the previous theories, the difficulty of accounting for the rapid rotations of the planets and the difficulty of accounting for their large angular momenta. The rotations of the planets are readily attributed to the collision and the angular momentum of the planets was derived from the shattered companion; it was there initially and the collision merely involved redistribution of the initial angular momentum amongst the fragments.

There has been a good deal of controversy about whether this theory of the origin of the solar system can be made to work. Very special conditions are required for the intruding star to be

able to get away, carrying along with it the bulk of
the disrupted companion of the Sun, whilst leaving
behind sufficient material to form the planets.
A complicating factor is that mathematics is not
able to trace out in detail exactly what would happen. It seems not impossible, however, that with
specially arranged conditions, in which not very
much latitude can be allowed in any particular, the
solar system might have been produced by a collision between an intruding star and a companion
to the Sun. But so many special assumptions are
involved that it has been remarked that the solar
system had a very narrow escape from never
coming into existence.

Though this is not a valid argument against the
theory, there are many who are disinclined to
accept a theory which seems so inherently improbable. This is not a logical attitude. Unless
we can find a more plausible origin for the solar
system, we must provisionally accept any theory
that can be made to work. If the theory is improbable we are able, at any rate, to infer something
about the likelihood that other stars may have
systems of planets circulating about them. This
will give us a clue whether life is likely to be widespread throughout the universe or to occur only
exceptionally.

On any theory that requires either a near
approach of a star to the Sun or an actual collision,
it must follow that planetary systems are of very
exceptional occurrence. The stars are so thinly
scattered through space that the chance of a close
approach of any two stars is extremely small.

The nearest known star is 25 million million miles away; it is obvious that, with stars at this sort of distance apart, it will be a very exceptional thing for one star to pass near or actually to collide with another. As it is difficult to conceive of distances measured by millions of millions of miles, we can get a picture that the mind can more readily grasp in this way. Suppose we have a hollow globe the size of the Earth, 8,000 miles in diameter, and that we put half a dozen tennis balls inside it and allow them to fly about in any direction, rebounding from the wall when they hit it. The chance that two of these balls will collide is about equal to the chance that two stars will come into collision.

The chance that two stars will actually collide was computed by Sir James Jeans. He found that a given star will be likely to meet with an actual collision only once in 600,000 million million (6×10^{17}) years. The chance of two stars approaching each other sufficiently closely, without an actual collision, to give rise to the ejection of a tidal filament is somewhat, but not much, greater.

From these figures we can make an estimate of the number of stars that are likely to have experienced a close approach to another star. As a rough, but sufficiently close estimate, we will suppose that any given star makes a close approach to another star on the average once in 500,000 billion [1] (5×10^{17}) years. The average age of the stars is believed to be greater than 10,000 million (10^{10}) years. It follows that not more than one star in fifty million is likely to have collided, or

[1] Billion is used to denote one million million.

nearly to have collided, with another star in the whole of its life. In our own stellar universe the number of stars is of the order of 100,000 million. These stars are not distributed uniformly; there are localised aggregations or clusterings where the star density is above the average and regions where the stars are relatively sparse. If we assume that the star density in the vicinity of the Sun is sufficiently representative of the system as a whole, it will follow that in our own stellar universe there are not likely to be more than several hundred stars that, at some time or other in the course of their life, have had a close approach to another star, which approach has resulted in the ejection of a tidal filament of matter.

No great exactitude can be claimed for these figures. But, after making adequate allowance for all possible uncertainties in the data, there is no escape from the conclusion that the number of stars in our stellar universe which in the course of their lifetime have made a close approach to another star is very small—a mere few hundred. Until recently it was thought that each such close approach would result in the formation of a planetary system; but now, as we have already mentioned, it is realised that this is far from being so and that it is only under very specialised assumptions that the theory could be made to work. The close approach will lead to the ejection of a stream of matter from the great tidal bulge drawn up on the star; but this matter will normally be dissipated away into space and will not condense into discrete planets, held bound to the parent Sun by the force of gravitation. To ensure this, we have been compelled to assume that the Sun was initially a double star; and not only so but, in order to make the theory workable, it becomes necessary to impose special conditions on the nature of the encounter between the two stars. There are limitations to the direction and velocity of approach of the intruding star which are necessary if a system like the solar system is to result from the encounter. The effect of these limitations is very seriously to reduce the chance that a close encounter between two stars will give birth to a system of planets. Precise estimation of the probability is not possible but it seems likely that out of several hundred encounters, each sufficiently close to result in the ejection of matter from the tidal protuberance, there can at the most be very few that give rise to planetary systems. Our solar system appears, therefore, to be a system that is nearly, though perhaps not quite, unique in our stellar universe.

There is one possible way of escape, however, from this conclusion. For the estimate of the chance of a close approach of two stars has been based upon the observed average distance apart of one star from another and it is implicitly assumed that this distance has remained the same throughout the lifetime of the stars. Modern investigations suggest, on the contrary, that the validity of this assumption is doubtful: that, in fact, the stars may have been very much closer together when the Earth and the other planets were born than they are at the present time.

It would lead us too far afield to go fully into the

reasons for this conjecture. There are, however, two main lines of approach. The first line of approach is through the consideration of the problem of the source of stellar energy. The age of the Earth is somewhere about 3,000 million years; the Sun must be at least as old as the Earth. And for this period of time it has been lavishly pouring out energy in the form of heat and light into space. Where does this energy come from? This is a problem that has occupied astronomers for many years. Lord Kelvin suggested that the Sun was contracting under its own gravitation; the process of contraction would release energy, which the Sun radiates away into space. It is now known that energy provided in this way would not maintain the Sun's radiation for more than about 20 million years. When radioactivity was discovered, it was thought that this might provide an explanation; but the supply of energy that could be provided again proved hopelessly inadequate.

Meanwhile, more precise information about the age of the Earth was being obtained and it was realised that the only processes that could maintain the Sun's radiation for thousands of millions of years must be of an atomic nature. There were two main alternatives. The energy might be derived from the actual annihilation of matter or from the building-up of heavier atoms from atoms of hydrogen. The modern conception of matter is that it is built up of elementary particles, protons and electrons, of positive and negative electricity. If we could bring a proton and an electron together, so that their charges coalesced and neutralised one another, both particles would disappear and a splash of energy would result. The amount of energy locked up in the atom, which might conceivably be released in this way, is prodigious. From one ounce of coal we should obtain sufficient energy to run engines of a total horse-power of 100,000 h.p. for one year; in other words, the fuel requirements of a large generating station could be satisfied by a fuel supply of one ounce of coal a year. If annihilation of matter provided the source of the energy of the stars, they would be able to maintain their output of energy without very serious diminution for millions of millions of years. For a time, indeed, various considerations, into the details of which it is not necessary to enter here, led astronomers to believe that the ages of the stars were, in fact, of the order of millions of millions of years. It seemed, therefore, that the process of annihilation of matter must actually be taking place within the intensely hot interiors of the stars.

There was a difficulty, however, which could not easily be resolved. Investigations into the internal constitution of the stars led to the conclusion that the temperatures at the centres of the stars were of the order of from 10 to 20 million degrees. Theoretical considerations suggested that the annihilation of matter could not occur at temperatures below some thousands of millions of degrees. The supposition that annihilation of matter can occur in the interior of the stars appears, therefore, to be untenable.

We are consequently thrown back on the alternative that the energy of the stars is derived from the building up of heavier atoms from atoms of hydrogen. A helium atom can be built up from four atoms of hydrogen. But the weight of the helium atom is less than the weight of the four hydrogen atoms by about 1 part in 140. From four pounds of hydrogen we could only obtain about 3 lb. $15\frac{1}{2}$ oz. of helium. What has happened to the missing half-ounce? It is accounted for by the energy that has been released in the process. We can suppose also that elements heavier than helium are built up from hydrogen atoms; the energy set free is then greater, though not appreciably greater, than in the building up of helium. The energy obtained by building up heavy atoms from other lighter atoms, with the exception of hydrogen, is also relatively small. It is the first step in the process of building up, the formation of helium from hydrogen, that provides the bulk of the energy that can be derived by the atom-building processes. It is conceivable that, in the beginning, all matter consisted of the elementary particles and that the heavy elements have been built up stage by stage from these particles. The process can go on until we arrive at the heavy atoms, such as those of radium, thorium and ceranium, which are inherently unstable and which spontaneously disintegrate.

The energy that can be released by atom-building processes is only about one per cent. of the energy that can be provided by the annihilation of matter. The time during which the stars can continue to radiate is therefore much shorter. The Sun contains at present about one-third part by weight of hydrogen; the other two-thirds consist of heavier elements. Whilst the age of the stars can be extended to several million million years if annihilation of matter takes place, the age is limited to about 10,000 million years if annihilation does not occur and atom-building provides the main source of energy. As the age of the Earth is about 3,000 million years, it would seem that the stars cannot be much older than the Earth itself, a conclusion that is somewhat unexpected.

The second line of approach to the question whether the stars were closer together when the planets were born than they are now is through the information that has been obtained about the expansion of the universe. It has been found by observation that the various stellar universes, the so-called spiral or extra-galactic nebulæ, are all receding from each other, with speeds that are proportional to their distances apart. From whichever of these universes observations are made, the other universes appear to be receding from it, and the more distant the universe the faster is its velocity of recession. This has been interpreted as evidence that the Cosmos as a whole is expanding. Imagine a rubber balloon being inflated and suppose that on its surface we have marked a number of ink-dots. As the balloon expands, each dot will increase its distance from every other dot and the rate of increase of the distance between any two dots will be proportional to that distance. This is analogous to what we observe in the motions of the various stellar universes.

16

The rate at which these universes are receding from each other is such that all distances are doubled in about 1,300 million years. If this rate of recession had remained constant in past time, it would follow that the various universes were very much closer together when the solar system was formed than they are now. If we go back somewhat farther in time, we should find that they were all crowded together in a relatively small volume of space; if our conclusions about the ages of the stars are correct, it was then that the stars were born.

When we seek to probe backwards in time, we do not seem to be able to get beyond a time a few thousand million years ago. At such a time in the past, we find the various universes congregated close together in a volume of space much smaller than they now occupy; at such a time the stars were born; at such a time the solar system itself was born. It seems as though time stood still until all sorts of things began to happen and time began to move. Exactly what did happen at that time, and why it happened, we do not know. It has been suggested that at this epoch, which measures for us the beginning of time, all the matter in the entire Cosmos was closely packed together—that there was in the beginning, in fact, one great Atom. Something then happened that was in the nature of an explosion; the Cosmos was shattered and fragments were hurtled in all directions through space. They would continue to move with their initial velocities and, at any subsequent time, the fastest-moving fragments would naturally be found to have moved to the greatest distances. The appearance would be exactly that revealed by observation, of an expanding universe. At about this same time, or perhaps shortly afterwards, the stars were born. In these early ages, when the universe as we know it was beginning to take shape, the stars must have been much closer together than they now are. Collisions and close approaches between stars, or between dense aggregations of stars, must have been frequent, and it is possible that in all this turmoil the birth of planetary systems may have been widespread.

A way of escape from the impasse into which we were landed in our attempts to account for the origin of the solar system may thus be provided. It had seemed that such a combination of initial conditions was required to enable the system to come into existence that the system must be almost unique; it now appears that the conditions under which the system was born may have been so different from those which are now revealed to us by observation that we cannot draw any certain conclusions. Though we are still unable to say in detail how the solar system originated, we are probably correct in concluding that it is by no means so exceptional as we had thought and that there are likely to be many other stars, in addition to the Sun, which ·are accompanied by systems of planets.

CHAPTER X

BEYOND THE SOLAR SYSTEM

IN the preceding chapter we discussed in some detail the origin of the solar system in the hope that we might be able to estimate the likelihood that other stars may have families of planets. We found that no definite answer was possible because we could not say with certainty how the solar system had come into being. It has proved to be one of the most difficult problems of astronomy. If the stars have always been distributed in much the same way as they are at present, we seem to be forced to the conclusion that the proportion of stars that have families of planets must be extremely small. We cannot be certain, of course, that we may not have overlooked some possibility. There seems to be no obvious loophole in the arguments, and I can see very little hope that the solar system will be explained by means of some event in the Sun's history that might equally well have happened to the majority of the stars. Provided that the distribution of the stars has not greatly changed in the past, it seems almost certain that very exceptional circumstances are required to explain the origin of the solar system.

Such was the position until recent years. The theory of the expanding universe was then formulated to provide an explanation of the observed motions of the distant universes, which appear to be all receding from us, the more distant the universe the greater being the velocity of recession. I use the word *appear* advisedly, because the theory of the expanding universe is based upon a particular interpretation of observations. If a body which is sending out radiations is moving towards us, the radiations are compressed and appear to us to be of shorter wave-length than they would if we were on the moving body; if the body is moving away from us, the radiations that we receive appear to be of longer wave-length. This change of wave-length due to motion is exemplified when a train passes with its whistle blowing, by the sudden drop in pitch of the note of the whistle as the train passes and instead of moving towards us is moving away.

The radiations that compose the light received from the distant universes are of longer wave-length than corresponding radiations from terrestrial sources. This is the observational fact which is interpreted as due to motions of recession of the distant universes. But is this necessarily the only interpretation? It has been suggested, for instance, that the light from these remote systems is in some way modified in the course of its long journey through space. Whether the observed changes in wave-lengths of the radiations from the remote systems are due to their motions or to a slow progressive change in wave-length as the light travels through space is a question that can ultimately be decided by observation. It is hoped that the great two-hundred-inch telescope, now in course of construction, will make a decision between the two alternatives possible; the great light-gathering power of this giant telescope, enabling it to explore space to much greater distances than any existing telescope, will be invaluable for this purpose.

If these observations prove that the conception of the expansion of the universe is not correct, we shall no longer have any reason to suppose that the average distances of the stars from one another were ever much different from what they now are. If, on the other hand, the observations prove that the universe as a whole is now in a state of expansion, what can we legitimately infer about its past history? To suppose that there has been a uniform expansion from an initial highly condensed condition is to make an assumption which, though it may possibly be correct, is without any justification from observation. The course of events in a changing universe can be discussed by mathematics, and it appears that there are various possibilities: a uniform expansion is one; pulsation, with alternate periods of expansion and contraction, is another. We have no means of deciding which course the universe has actually followed. It may be true, or it may not, that the average distance apart of the stars was at one time much less than it is at present; planetary systems may be relatively few in number or, on the other hand, they may be far more numerous than was believed until recently.

Whichever view we incline to take, however, we are certainly not justified in supposing that the solar system is unique. It has somehow come into existence, and it is not logical to suppose that other systems could not come into existence in a similar way. The probability might be extremely small, and yet the number of planetary systems in the whole universe could be considerable, because the number of stars in each of the separate stellar universes and the number of these universes are both very great. It has been estimated that there are about 100 million universes in the region of space that can be probed by the one-hundred-inch telescope; if the number of stars with a family of planets did not average more than one per universe, the total number of planetary systems would still be considerable. It seems to me that we cannot avoid the conclusion that the total number of planetary systems must be very large.

But the existence of other planetary systems, though a necessary condition for life to exist else-where in the universe, is not a sufficient condition. In any planetary system everything seems to be weighted against the possibility of the existence of life; a somewhat precise adjustment of conditions is needed in order that life may be possible. If the planet is very near its parent Sun, it will be too hot for life to exist; if it is very far away, it will be too cold. If it is very much smaller than the Earth, it will have been unable to retain any atmosphere. If it is much larger, it will have retained too much atmosphere; for when the gravitational attraction is so great that hydrogen cannot escape from the atmosphere the formation of the poisonous gases, ammonia and marsh-gas—which we found in the atmosphere of Jupiter and Saturn—appears to be almost inevitable. There seems to be little chance that life can exist on any world if that world differs greatly from the Earth in size and weight; it must be neither very much smaller than the Earth nor very much larger.

But this is not the only restriction. The stars differ enormously in candle-power or luminosity. There are some stars, called giant stars, whose luminosity is many thousands of times greater than that of the Sun. There are other stars, called dwarf stars, whose luminosity is very much smaller than that of the Sun. Thus, for instance, Canopus is about eighty thousand times more luminous than the Sun; the Sun, on the other hand, is about sixteen thousand times more luminous than the faint companion of the bright star Procyon. If the Sun were replaced by Canopus, the Earth would become so intensely hot that all life, both plant life and animal life, would cease at once. The surface of the Earth would be seared as though by the blast from a furnace and the oceans would be rapidly vaporised. If the Sun were replaced by the companion of Procyon, the Earth would receive so little warmth that all the oceans would be frozen and the cold would be so intense that life would again be out of the question. Thus not only are there somewhat narrow limits to the size of a planet if it is to be the home of life, but also for any given parent Sun there are somewhat narrow limits of distance within which the planet must be situated in order that it may be neither too hot nor too cold for life.

In the case of the solar system, if the Earth were as near the Sun as Mercury is, it would be too hot for any life to be possible on it; if, on the other hand, it were as far from the Sun as

Jupiter, it would be much too cold for life to be possible.

A somewhat precise adjustment of two factors, the size of the planet and its distance from its parent Sun, thus seems to be essential if life is to be possible on the planet. It is not enough to have either factor satisfied without the other. There may be many planetary systems entirely devoid of life, either because the planets are too large or because they are too small or because they are too near the parent Sun or are too remote from it. It is not possible to make any estimate of the proportion of planets that will fall within the appropriate limits of size and of distance from the parent Sun, though the proportion is likely to be small.

To sum up the argument: the conditions needed for birth to be given to a planetary system may be so exceptional that amongst the vast number of stars in any one stellar universe we may expect to find only a very limited number that have a family of planets; and amongst these families of planets there cannot be more than a small proportion where the conditions are suitable for life to exist. Life elsewhere in the universe is therefore the exception and not the rule. If we could travel through the universe and survey each star in turn, we should not find life here, there and everywhere. Occasionally in our wanderings we should find a star with a family of planets; few of these could be the home of life, but some there would be which would comply with our requirements. If the proportion of planets on which life can exist is not more than one in a thousand, or even one in a million, the total number of worlds that are suitable for life would yet be considerable, so vast is the scale on which the universe is constructed.

It may still be objected that even where conditions are suitable for life to exist there may be no life. We cannot hope ever to have any direct information about these remote worlds; we can only be guided by what we have learnt from the study of worlds near at hand. The evidence that there is vegetation on Mars is almost conclusive, and affords very strong presumptive evidence that life will appear when conditions are suitable for it.

It is idle to try to guess what forms life might take in other worlds. The human mind cannot refrain from toying with the idea that somewhere in the universe there may be intelligent beings who are the equals of Man, or perhaps his superiors; beings, we may hope, who have managed their affairs better than Man has managed his. Neither the investigations of the astronomer nor the investigations of the biologist can help us in this matter. It must remain for ever a sealed book. But it is unlikely that evolution has followed a parallel course on any two worlds. Small differences in conditions, in climate, in temperature, in atmosphere and in topography, may prove of fundamental importance. As evolution proceeds, it may well be that here and there it branches off into this or that direction, when some slight change of conditions—perhaps trivial in itself—might have resulted in a branching off into some different direction, changing entirely the whole subsequent development.

There is one further point to mention, the question of the time scale. If we could move not merely through space but backwards and forwards in time, we should probably find life appearing, passing through its sequence of evolution, reaching its zenith of attainment and then passing away on one world after another. Our study of the Sun's family of planets suggested that Mars is a world long past its prime and that Venus is a world that is perhaps destined to be the home of life in the future. The zenith of life will not occur at the same time on all the worlds that are capable of supporting life: it may be that life in its most highly developed stages will not persist on any particular world for more than a small fraction of its life-history.

So we come to the end of our quest. We have seen that throughout the whole universe there is an essential uniformity in the structure of matter. The same few elements occur everywhere, to the remotest parts of the universe that we have been able to study. They are the bricks from which all matter—from its simplest to its most complex forms —are built up. We have seen also that matter obeys the same laws throughout the universe. The chemistry of the carbon atom is of fundamental importance in building up the great variety of the extremely complex molecules that form the basis of living matter. The carbon atom must play the same key part wherever in the universe life may be found. This leads us naturally to the conclusion that living matter is possible only under somewhat specialised and restricted conditions. These restrictions serve as a guide in estimating the likelihood that life may be found to occur on this world or on that world.

We have assumed that if the required conditions for life to be possible are obtained life will automatically make its appearance. This assumption may be criticised as unjustifiable in the absence of any knowledge as to how life originated on the Earth. Yet it appears more plausible than any other assumption that we can make and receives some confirmation from the fact that we have direct evidence of life—though admittedly only of plant life—on Mars, which is the only planet where we might expect to find any evidence of life at the present time.

But life is not widespread in the universe. The blazing suns, which we call stars, are far too hot to permit of the existence of any but the simplest chemical compounds. It is only on the cooler planetary bodies that we can hope to find life. Life is not possible, however, on the large majority of the planets: some have no atmospheres, others have atmospheres that are poisonous; some are too hot, others are too cold. And not more than a small proportion of the stars are likely to have any planets at all. With the usual prodigality of Nature, the stars are scattered far and wide, but only the favoured few have planets that are capable of supporting life. The radiations from millions upon millions of stars are being sent out, and for thousands of millions of years have been sent out, into space. An almost insignificant fraction of the radiations from a star here and a star there is utilised in making life possible on an attendant planet; the great bulk of the radiation is merely destined to travel on endlessly through space.

Yet though these restrictions severely winnow down the possible abodes of life in the universe, we cannot resist the conclusion that life, though rare, is scattered throughout the universe. It may be compared to a rare plant, whose distribution is widespread, but of which never more than a single specimen is found at a time. If life is the supreme purpose of creation, it may be a matter for some surprise that its occurrence is so restricted. It might have been expected that every star would minister to life; it has, in fact, often been asserted that this is so. But astronomy gives no support to this view. The task of the astronomer is to learn what he can about the universe as he finds it. To endeavour to understand the purpose behind it and to explain why the universe is built as it is, rather than on some different pattern which might have accorded better with our expectations, is a more difficult task; for this the astronomer is no better qualified than anybody else.

INDEX